Bioanalysis of Pharmaceuticals

Bioanalysis of Pharmaceuticals

Sample Preparation, Separation Techniques, and Mass Spectrometry

Editors

STEEN HONORÉ HANSEN

School of Pharmaceutical Sciences, University of Copenhagen, Denmark

STIG PEDERSEN-BJERGAARD

School of Pharmacy, University of Oslo, Norway
School of Pharmaceutical Sciences, University of Copenhagen, Denmark

WILEY

This edition first published 2015
© 2015 John Wiley & Sons, Ltd

Registered office
John Wiley & Sons Ltd, The Atrium, Southern Gate, Chichester, West Sussex, PO19 8SQ, United Kingdom

For details of our global editorial offices, for customer services and for information about how to apply for permission to reuse the copyright material in this book please see our website at www.wiley.com.

Library of Congress Cataloging-in-Publication Data applied for.

A catalogue record for this book is available from the British Library.

ISBN: 9781118716823

Set in 10/12pt TimesLTStd by SPi Global, Chennai, India

Contents

Contributing Authors

Leon Reubsaet
School of Pharmacy, University of Oslo, Norway
Chapters 2, 7, and 10

Trine Grønhaug Halvorsen
School of Pharmacy, University of Oslo, Norway
Chapters 6 and 10

Astrid Gjelstad
School of Pharmacy, University of Oslo, Norway
Chapter 6

Martin Jørgensen
Drug ADME Research, H. Lundbeck AS, Denmark
Chapter 11

Morten A. Kall
Department of Bioanalysis, H. Lundbeck AS, Denmark
Chapter 11

Preface

The field of bioanalysis is very broad, complex, and challenging, and therefore writing an introductory textbook in this field is a difficult task. From our point of view, a good introductory student textbook is limited in the number of pages, discusses the different principles and concepts clearly and comprehensively, and contains many relevant and educational examples. Given these criteria, we have narrowed our focus on bioanalysis. First, we have limited our discussion to the chemical analysis of *pharmaceuticals* that are present in biological fluids. The focus is directed toward substances that are administered as human drugs, including low-molecular drug substances, peptides, and proteins. Endogenous substances are not discussed. Second, the discussion of different analytical methods has been limited to those based on *chromatography* and *mass spectrometry*. Certainly, different immunological methods are also used, but teaching all the principles and applications of chromatographic, mass spectrometric, and immunological methods was too ambitious to meet our criteria for a good introductory student textbook.

The present book is the first introductory student textbook on chromatography and mass spectrometry of pharmaceuticals present in biological fluids, highlighting an educational presentation of the principles, concepts, and applications. We discuss the chemical structures and properties of low- and high-molecular pharmaceuticals, the different types of biological samples and fluids that are used, how to prepare the samples by extraction, and how to perform the final analytical measurement by use of chromatography and mass spectrometry. Many examples illustrate the theory and applications, and the examples discuss all practical aspects, including the calculations. Thus, in this textbook, you will even learn how to convert the numbers recorded by the instrument to the concentration of the actual drug substances in the biological sample.

Bioanalysis is an applied scientific discipline, and this represents another challenge in terms of writing an introductory student textbook. University professors are well trained in teaching the basic principles. However, bioanalysis is mainly performed outside the university by researchers in the pharmaceutical industry, in contract laboratories, and in hospital laboratories. Thus, the researchers outside the university have the best overview of the most important applications and techniques in practical use. To address this, both university professors and researchers from the pharmaceutical industry have authored this textbook. Hopefully, this has resulted in a textbook that reflects bioanalysis in the year 2015. The authors have been in close contact with colleagues for advice, and we would especially like to thank Elisabeth Leere Øiestad for fruitful discussions.

The present textbook is intended for the fourth- or fifth-year university pharmacy or chemistry student. Reading the textbook requires basic knowledge in organic chemistry and biochemistry, as well as in analytical chemistry. With respect to the latter, we have given priority to discuss the analytical techniques in a fundamental and educational frame, and detailed knowledge on instrumental analytical methods is not required prior to reading this textbook.

Good luck with the reading!

Oslo and Copenhagen, June 2014
Steen Honoré Hansen, Stig Pedersen-Bjergaard, Leon Reubsaet, Astrid Gjelstad,
Trine Grønhaug Halvorsen, Martin Jørgensen, and Morten A. Kall

 Readers can access PowerPoint slides of all figures at http://booksupport.wiley.com

1

Introduction

Stig Pedersen-Bjergaard

School of Pharmacy, University of Oslo, Norway
School of Pharmaceutical Sciences, University of Copenhagen, Denmark

Welcome to the field of bioanalysis! Through reading of this textbook, we hope you get fascinated by the world of bioanalysis, and also we hope that you learn to understand that bioanalysis is a highly important scientific discipline. In this chapter, five fundamental questions are raised and briefly discussed as an introduction to the textbook: (i) What is bioanalysis? (ii) What is the purpose of bioanalysis? (iii) Where is bioanalysis conducted? (iv) Why do you need theoretical understanding and skills in bioanalysis? And (v) how do you gain the understanding and the skills from reading this textbook?

1.1 What Is Bioanalysis?

In this textbook, we define *bioanalysis* as the chemical analysis of pharmaceutical substances in biological samples. The purpose of the chemical analysis is normally both to *identify* (identification) and to *quantify* (quantification) the pharmaceutical substance of interest in a given biological sample. This is performed by a *bioanalytical chemist* (scientist) using a *bioanalytical method*. The pharmaceutical substance of interest is often termed the *analyte*, and this term will be used throughout the textbook. Identification of the analyte implies that the exact chemical identity of the analyte is established unequivocally. Quantification of the analyte implies that the concentration of the analyte in the biological sample is measured. It is important to emphasize that quantification is associated with small inaccuracies, and the result is prone to errors. Thus, the quantitative data should be considered as an estimate of the true concentration. Based on theoretical and practical skills, and based on careful optimization and testing of the bioanalytical methods, the bioanalytical chemist tries to reduce the error level, providing concentration estimates that are very close to the true values.

Bioanalysis of Pharmaceuticals: Sample Preparation, Separation Techniques and Mass Spectrometry,
First Edition. Steen Honoré Hansen and Stig Pedersen-Bjergaard.
© 2015 John Wiley & Sons, Ltd. Published 2015 by John Wiley & Sons, Ltd.

Bioanalytical data are highly important in many aspects. As an example, a patient serum sample is analyzed for the antibiotic drug substance gentamicin, and gentamicin is measured in the sample at a concentration of 5 µg/ml. First, the identification of gentamicin in the blood serum sample confirms that the patient has taken the drug. This is important information because not all patients actually comply with the prescribed medication. Second, the exact concentration of gentamicin measured in the blood serum sample confirms that the amount of gentamicin taken is appropriate, as the recommended concentration level should be in the range of 4–10 µg/ml. For aminoglycoside antibiotics such as gentamicin, it is recommended to monitor the concentration in blood if the treatment is expected to continue for more than 72 hours as these antibiotics have the potential to cause severe adverse reactions, such as nephrotoxicity and ototoxicity.

As will be discussed in much more detail in this book, not only blood serum samples are used for bioanalysis. Bioanalysis can be performed on raw blood samples (whole blood) or on blood samples from which the blood cells have been removed (serum or plasma). Alternatively, bioanalysis can be performed from urine or saliva as examples, depending on the purpose of the bioanalysis. Bioanalysis is performed both on human samples and on samples from animal experiments.

1.2   What Is the Purpose of Bioanalysis, and Where Is It Conducted?

Bioanalysis is conducted in the *pharmaceutical industry*, in *contract laboratories* associated with the pharmaceutical industry, in *hospital laboratories*, in *forensic toxicology laboratories*, and in *doping control laboratories*. In the pharmaceutical industry and in the associated contract laboratories, bioanalysis is basically conducted to support the development of new drugs and new drug formulations. In hospital laboratories, bioanalysis is used to monitor existing drugs in patient samples, to check that individual patients take their drugs correctly. In forensic toxicology laboratories and doping laboratories, bioanalysis is used to check for abuse of drugs and drug-related substances.

1.2.1   Bioanalysis in the Pharmaceutical Industry

Bioanalytical laboratories are highly important in the development of new drugs and new drug formulations in the pharmaceutical industry. Thus, identification and quantification of drug substances and metabolites in biological samples like blood plasma, urine, and tissue play a very important role during drug development. Drug development begins with the identification of a medical need and hypotheses on how therapy can be improved. *Drug discovery* is the identification of new *drug candidates* based on combinatorial chemistry, high-throughput screening, genomics, and ADME (absorption, distribution, metabolism, and elimination). By combinatorial chemistry, a great number of new drug candidates are synthesized, and these are tested for pharmacological activity and potency in high-throughput screening (HTS) systems. The HTS systems simulate the interaction of the drug candidates with a specific biological receptor or target. Once a *lead compound* is found, a narrow range of similar drug candidates is synthesized and screened to improve the activity toward the specific target. Other studies investigate the ADME profile of drug candidates by analyzing samples collected at different time points from dosed laboratory animals (*in vivo testing*) and tissue cultures (*in vitro testing*).

Drug candidates passing the discovery phase are subjected to toxicity testing and further metabolism and pharmacological studies in the *preclinical development* phase. Both in vivo and in vitro tests are conducted, and various animal species are used to prove the pharmacokinetic profile of the candidate. The detailed information about the candidate forms the basis for further pharmaceutical research on the synthesis of raw materials, the development of dosage forms, quality control, and stability testing.

The *clinical development* phase can begin when a regulatory body has judged a drug candidate to be effective and to appear safe in healthy volunteers. In *phase I*, the goal is to establish a safe and efficient dosage regimen and to assess pharmacokinetics. Blood samples are collected and analyzed from a small group of healthy volunteers (20–80 persons). The data obtained form the basis for developing controlled *phase II* studies. The goal of phase II studies is to demonstrate a positive benefit–risk balance in a larger group of patients (200–800) and to further study pharmacokinetics. Monitoring of efficacy and monitoring of possible side effects are essential. Phase II studies can take up to two years to fulfill. At the end of phase II, a report is submitted to the regulatory body, and conditions for phase III studies are discussed. Additional information supporting the claims for a new drug is provided. *Phase III* begins when evidence for the efficacy of the drug candidate and supporting data have demonstrated a favorable outcome to the regulatory body. The phase III studies are large-scale efficacy studies with focus on the effectiveness and safety of the drug candidate in a large group of patients. In most cases, the drug candidate is compared with another drug already in use for treatment of the same condition. Phase III studies can last two to three years or more, and 3000–5000 patients can be involved. Carcinogenetic tests, toxicology tests, and metabolic studies in laboratory animals are conducted in parallel. The cumulative data form the basis for filing a new drug application to the regulatory body and for future plans for manufacturing and marketing. The regulatory body thoroughly evaluates the documentation that is provided before a market approval can be authorized and the drug product can be legally marketed. The time required from drug discovery to product launch is up to 12 years. Phase IV studies are studies that are conducted after product launch to demonstrate long-term effects and new claims, expand on approved claims, examine possible drug–drug interactions, and further assess pharmacokinetics. Several thousand patients participate in phase IV studies.

Bioanalytical measurements are conducted during drug discovery, preclinical development, and clinical development, and they are intended to (among other things) generate the experimental data to establish the *pharmacokinetics*, the *toxicokinetics*, and the *exposure–response* relationships for a new drug. The pharmacokinetics of a certain drug substance describes how the body affects the drug after administration (ADME): how the drug is absorbed (A) and distributed (D) in the body, and how the drug is metabolized (M) by metabolic enzymes and chemically changed to different types of metabolites, which in turn are excreted (E) from the body. Bioanalysis is used extensively in pharmacokinetic studies, among others, to establish blood concentration–time profiles, and to measure the rate of drug metabolism and excretion. This involves a large number of both animal and human samples.

Toxicokinetics studies, in contrast, are intended to investigate the relationship between the exposure of a new drug candidate in experimental animals and its toxicity. This type of information is used to establish a relationship between the possible toxic properties of a drug in animals and those in humans. Toxicokinetic studies involve bioanalysis in both animal and human samples.

Exposure–response studies investigate the link between pharmacokinetics and pharmaco-dynamics. *Pharmacodynamics* is the study of the biochemical and physiological effects of a drug substance on the body. Exposure–response studies thus establish the link between the dose, the blood concentration, and the effect. Also for exposure–response studies, bioanal-ysis of a large number of blood samples has to be conducted. Frequently, such bioanalytical measurements are outsourced by the pharmaceutical company to a contract laboratory that is highly specialized in bioanalysis. High quality of the bioanalytical data is mandatory, because these data from drug discovery, preclinical, and clinical studies are used to support the regulatory filings.

1.2.2   Bioanalysis in Hospital Laboratories

Bioanalysis is also very important in many hospital laboratories. Here, the focus is on mea-suring the drug concentration in blood samples of patients to check that they are properly medicated. This is called *therapeutic drug monitoring* (TDM), and it refers to the individ-ualization of dosage by maintaining serum or plasma drug concentrations within a target range to optimize efficacy and to reduce the risk of adverse side effects. The target range of a drug is also called the *therapeutic range* or the *therapeutic window*; it is the concentration range between the lowest drug concentration that has a positive effect and the concentra-tion that gives more adverse effects than positive effects. Variability in the dose–response relationship between individual patients is due to *pharmacokinetic variability* and *phar-macodynamic variability*, as shown in Figure 1.1.

Pharmacodynamic variability arises from variations in drug concentrations at the recep-tor and from variations in the drug–receptor interaction. Pharmacokinetic variability is due to variations in the dose to plasma concentration relationship. Major sources of pharma-cokinetic variability are age, physiology, disease, compliance, and genetic polymorphism of drug metabolism. Indications for including a drug in a therapeutic drug-monitoring pro-gram are:

- There is an experimentally determined relationship between the plasma drug concentra-tion and the pharmacological effect.
- There is a narrow therapeutic window.
- The toxicity or lack of effectiveness of the drug puts the patient at risk.
- There are potential patient compliance problems.
- The dose cannot be optimized by clinical observations alone.

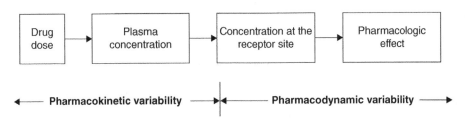

Figure 1.1 *Effects of pharmacokinetics and pharmacodynamics on the dose–response relationship*

Table 1.1 *Therapeutic range of common drugs subjected to therapeutic drug monitoring*

Drug	Therapeutic range	Drug	Therapeutic range
Amitriptyline	120–150 ng/ml	Nortriptyline	50–150 ng/ml
Carbamazepine	4–12 µg/ml	Phenobarbital	10–40 µg/ml
Desipramine	150–300 ng/ml	Phenytoin	10–20 µg/ml
Digoxine	0.8–2.0 ng/ml	Primidone	5–12 µg/ml
Disopyramide	2–5 µg/ml	Theophylline	10–20 µg/ml
Ethosuximide	40–100 µg/ml	Valproic acid	50–100 µg/ml
Lithium	4–8 µg/ml		

Many drugs do not meet the criteria to be included in a TDM program. They are safely taken without determining drug concentrations in plasma because the therapeutic effect can be evaluated by other means. For example, the coagulation time effectively measures the efficacy of an anticoagulant drug, and the blood pressure indicates the efficacy of a drug used in the treatment of hypertension. In these situations, it is preferred to adjust the dosage on the basis of medical response.

The two major situations when TDM is advised are (1) for drugs used prophylactically to maintain the absence of a condition (e.g., depressive or manic episodes, seizures, cardiac arrhythmias, organ rejection, and asthma relapses) and (2) to avoid serious toxicity for drugs with a narrow therapeutic window (e.g., antiepileptic drugs, antidepressant drugs, digoxin, phenytoin, theophylline, cyclosporine and HIV protease inhibitors, and amino-glycoside antibiotics). The therapeutic range of some drugs subjected to TDM is shown in Table 1.1.

1.2.3 Bioanalysis in Forensic Toxicology Laboratories

Bioanalysis is a core discipline also in forensic toxicology laboratories, where a large number of blood, urine, and saliva samples are analyzed to identify abuse of drugs and narcotics. The focus is on drugs and their metabolites, narcotics, and other substances that are toxicologically relevant. Serious cases that are of criminal relevance may include:

- Analysis of pharmaceuticals and addictive drugs that may impair human behavior.
- Detection of poisons and evaluation of their relevance in determining causes of death.

In forensic toxicology, the analyte is essentially unknown. Therefore, samples are first screened for the presence of drugs or drugs of abuse. In case of a positive sample, the drug or the drug of abuse is confirmed with a second bioanalytical method. Due to the serious legal consequences of forensic cases, particular emphasis is placed on the quality and reliability of bioanalytical results. The work always involves the application of at least two different analytical methods (screening and confirmation) based on different physical or chemical principles.

1.2.4 Bioanalysis in Doping Control Laboratories

Bioanalysis also is very important in doping control laboratories, where blood and urine samples are tested for doping agents. Only laboratories accredited by the World

Anti-Doping Agency (WADA) take part in the testing. WADA was established in 1999 as an international agency to promote, coordinate, and monitor the fight against doping in sport. One of WADA's most significant achievements was the acceptance and implementation of the World Anti-Doping Code (the Code). The Code is the core document that provides the framework for antidoping policies, rules, and regulations within sport organizations and among public authorities. The Code works in conjunction with five international standards aimed at bringing harmonization among antidoping organization in various areas. The standards are:

- List of Prohibited Substances and Methods
- International Standard for Testing
- International Standard for Laboratories
- International Standard for Therapeutic Use Exemptions
- International Standard for the Protection of Privacy and Personal Information.

The prohibited list is the standard that defines substances and methods that are prohibited to athletes at all times (both in competition and out of competition), substances prohibited in competition, and substances prohibited in particular sports. The prohibited list is updated annually.

The purpose of the International Standard for Testing is to plan for effective in-competition and out-of-competition testing and to maintain the integrity and identity of the samples collected. The International Standard for Therapeutic Use Exemptions and the International Standard for the Protection of Privacy and Personal Information ensure that the process of granting an athlete therapeutic-use exemptions is harmonized and that all relevant parties adhere to the same set of privacy protections.

The purpose of the International Standard for Laboratories is to ensure that laboratories produce valid test results. The standard further ensures that uniform and harmonized results are reported from all accredited laboratories. In addition, the document specifies the criteria that must be fulfilled by antidoping laboratories to achieve and maintain their WADA accreditation.

1.3   Bioanalysis Is Challenging

Bioanalysis is highly challenging because most target pharmaceutical substances are present in blood, urine, and saliva samples at very low concentrations. Typically, the concentration level is at the low ng/ml level, but in some cases, target pharmaceuticals have to be detected even down to the pg/ml level. This relies on very sensitive instrumentation and high operator skills. In addition, the target pharmaceuticals coexist with a broad range of endogenous compounds that are naturally present in biological samples. There can be thousands of different components, and many of them can be present at high concentration levels. Therefore, in most cases, a successful bioanalysis procedure requires the isolation of target pharmaceuticals from the biological matrix, before the final measurement with a sensitive instrument. Thus, experience and skills on how to prepare samples are extremely important in bioanalysis. The intention of the current textbook is to provide the reader with the required theoretical understanding and skills related to the understanding, development, and application of bioanalytical methods and procedures.

1.4   The Different Sections of This Textbook

The first part of this book is focused on the chemical properties of drug substances (Chapter 2) and the properties of the different biological fluids in use (Chapter 3). Careful reading of Chapter 2 is important for readers who are not familiar with pharmaceutical substances or the chemical properties of these substances. Understanding the chemical properties of the target pharmaceuticals is highly important in order to understand the bioanalytical procedures. Chapter 3 discusses the properties of different biological fluids, and if you are unfamiliar with biological fluids, you should read this chapter carefully. Understanding the properties of the biological fluids is mandatory in order to understand bioanalytical procedures.

Chapters 4–8 teach the different techniques and their principles, with foci on sample preparation, separation, and detection. These are chapters that are similar to the content of general textbooks in analytical chemistry. So, if you have been through general courses in analytical chemistry, this part of the textbook will be repetition. The discussion about sample preparation gives the reader an understanding of how to isolate target pharmaceuticals from the bulk biological matrix. The discussion about separation is focused on chromatographic separation, in which target pharmaceuticals are separated from any other substances in the sample, and the discussion about detection is focused on the final measurement of the substance, in most cases by mass spectrometry.

Chapters 9 and 10 are a collection of examples of bioanalytical procedures that are typically not found in general textbooks in analytical chemistry. In this part of the textbook, we make use of all previous knowledge and try to give you the full understanding. In these chapters, we discuss practical examples, and all discussions are related to the theory presented in Chapters 4–8. Remember when you read Chapters 9 and 10 that you should understand rather than remember all the details. Hopefully, the combination of Chapters 4–10 should give the reader a very good understanding of bioanalytical procedures. Finally, regulatory aspects related to bioanalytical procedures are discussed in Chapter 11. This is important to make sure that we can rely on the data generated from the application of bioanalytical methods.

Good luck with your journey into the world of bioanalysis!

2

Physicochemical Properties of Drug Substances

Steen Honoré Hansen[1] and Leon Reubsaet[2]

[1]*School of Pharmaceutical Sciences, University of Copenhagen, Denmark*
[2]*School of Pharmacy, University of Oslo, Norway*

The development of both sample preparation strategies and chromatographic methods is based on the physicochemical properties of the substances to be analyzed as well as the principles of the analytical technique used. Why is it that in some cases mass spectrometric detection is needed to determine a substance, whereas in other cases UV detection is sufficient? Can chromatographic behavior be predicted from simply looking at the chemical structure of the analyte? In this chapter, the most important physicochemical properties of small-molecule drug substances as well as those of peptide and protein biopharmaceuticals are discussed. The discussions are short and comprehensive, as most of this information should already be known from learning general chemistry. The properties discussed here will be used in other chapters in this book in relation to sample preparation and subsequent analysis.

2.1 Bioanalysis in General

In bioanalysis, the task often is to perform *qualitative or quantitative measurements* of analytes in complex matrices consisting of thousands of other chemical entities. Therefore, a high degree of selectivity is needed to be able to "pick the needle out of the haystack" and in this way increase the *reliability* of the data obtained. In many bioanalytical methods, the *selectivity* is incorporated at several stages: in the sample preparation, in the following chromatographic separation, and in the detection step. To be able to optimize the selectivity, a basic knowledge of some fundamental chemical and *physicochemical properties* is needed.

Bioanalysis of Pharmaceuticals: Sample Preparation, Separation Techniques and Mass Spectrometry,
First Edition. Steen Honoré Hansen and Stig Pedersen-Bjergaard.
© 2015 John Wiley & Sons, Ltd. Published 2015 by John Wiley & Sons, Ltd.

2.2   Protolytic Properties of Analytes

Many small molecules show *protolytic properties*, which cause them to be present in an ionic state as well as a neutral form. The degree of ionization controlled by the surrounding aqueous solvent very much influences the properties of the molecules and thus their behavior in each step in the bioanalytical method.

pH is defined in dilute aqueous solution and is an expression for the acidity or alkalinity of an aqueous solution. The pH concept is extremely important and has great influence on living organisms as well as in analytical chemistry. Water can react with itself to form a hydronium ion and a hydroxide ion:

$$2H_2O \rightleftarrows H_3O^+ + OH^- \tag{2.1}$$

This is called *autoprotolysis*, as the water in this case acts as an acid and a base. The autoprotolysis constant is:

$$K_w = \frac{[H_3O^+][OH^-]}{[H_2O]^2} = 10^{-14} \tag{2.2}$$

and it indicates that only a very small amount of water is ionized. The concentration of the two ions, H_3O^+ and OH^-, in pure water is therefore 10^{-7} M of each ion.

pH is defined as the negative logarithm to the activity, $a_H{}^+$, or the concentration of the hydrogen ion, $[H^+]$ (being equivalent to the hydronium ions):

$$pH = -\log(a_{H^+}) \approx -\log([H^+]) \tag{2.3}$$

Strong acids and strong bases are fully ionized in dilute aqueous solution, and the activity and concentration of $[H^+]$ therefore can be considered to be identical.

Weak acids and weak bases are not completely ionized in aqueous solution and are therefore in equilibrium with the unionized acid or base. When we ignore the weak autoprotolysis of water, we get the following general equation for a weak acid:

$$HA \rightleftarrows H^+ + A^- \Rightarrow K_a = \frac{[H^+][A^-]}{[HA]} \tag{2.4}$$

When an acid (H^+) is added to such a system, the H^+ will partly be removed by association with A^- to form HA, and if a base (OH^-) is added it will be partly neutralized by H^+ and more HA will dissociate. pH will thus be maintained in the solution. A system like this is called a *buffer system*, and the purpose of a buffer is to maintain the pH in the solution. *pK*$_a$ is defined as the negative logarithm to K_a:

$$pK_a = -\log K_a \tag{2.5}$$

and it is obvious that the highest *buffer capacity* is achieved at a pH value equivalent to the pK_a value of the buffer substance. Combining Equations 2.4 and 2.5 results in a most useful equation called the *Henderson–Hasselbalch equation*:

$$pH = pK_a + \log\frac{[A^-]}{[HA]} \tag{2.6}$$

At $pH = pK_a$, equal concentrations of the acid and corresponding base are present. If the ratio between HA and A^- becomes 9/1 (only 10% base) pH will decrease one unit, and if the ratio becomes 99/1 (1% base) the pH value will decrease by two units. This is illustrated

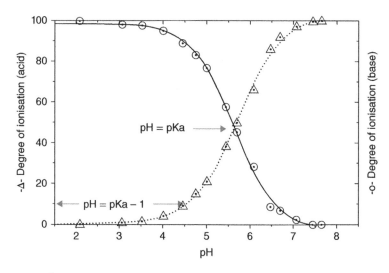

Figure 2.1   *Ionization of acids and bases as a function of pH*

in Figure 2.1. Equivalent estimations can be performed when increasing the base content. One example of a buffer system is arterial plasma, and this is featured in Box 2.1.

Box 2.1 Arterial Plasma Is a Buffer System

pH in arterial plasma is buffered by a special bicarbonate system. When acid is added CO_2 is formed, which is actively controlled by the lungs, and if base is added the

Table 2.1   *Typical pK_a values of important functional groups*

Functional group	pK_a	Comments (depending on chemical structure)
R-COOH, carboxylic acid	4–5	Can be lower (more acidic)[a]
R-NH$_2$; R1, R2, NH; and R1, R2, R3, N, aliphatic amines	8–11	Can be lower (less basic)[a]
Aromatic amines	About 5	—
Quaternary ammonium ions	—	Ions with no protolytic properties; are always positively charged
Ar-OH, phenols	8–10	Can be lower (more acidic)[a]
R-OH, alcohols	>12	Can for practical purposes be considered as neutral substances
R-SO$_2$OH, sulfonic acid	About 1	Are for all practical purposes always negatively charged
R-CO—NH—CO—R and R—SO$_2$NH—R	7–11	Weak to very weak acids

[a] This depends on other chemical groups in the molecule.

increase in bicarbonate is actively controlled by the kidneys. In this way it is possible to maintain a pH of 7.4, although the pK_a value of the bicarbonate is 6.1. Plasma collected for bioanalysis no longer has the contact to the lungs and the kidneys, but the plasma still has some buffer capacity, mainly due to the content of about 8% of proteins.

It is convenient to have general knowledge of the pK_a values of a number of functional groups, as presented in Table 2.1.

The *pK_a value for bases* refers to the protonated form of the bases. However, the basicity of bases may also be expressed equivalent to the pK_a of acids. In that case, the term pK_b is used and

$$pK_a + pK_b = 14. \tag{2.7}$$

2.3   Partitioning of Substances

A prerequisite in chromatography as well as in many sample preparation techniques is the partitioning of molecules between more or less immiscible phases (gas–liquid, gas–solid, liquid–liquid, or liquid–solid). When molecules are in solution, they will be exposed to a number of *intermolecular interactions*. These include, among other things, diffusion, collisions, dipole–dipole interactions, hydrogen bonding, and electrostatic interactions, as illustrated in Table 2.2. The nature of the interactions taking place is dependent on the physical and chemical nature of the analytes, and these interactions will determine how the molecules are distributed between different phases.

Ionic interactions can be as strong as a covalent bond but are often limited to one interaction per molecule. In contrast, *van der Waal interactions* are relatively weak but have many interactions per molecule and therefore are also very important.

The partition or distribution between phases (see Figure 2.2) is also influenced by pH, and thus a thorough knowledge of the pH concept, including pK_a, as well as of distribution constants will ease the development of bioanalytical methods (e.g., the chromatographic separation). The distribution is dependent on the nature of the two phases as well as the temperature. If we want to alter the partition between the two phases, we must change one

Table 2.2   Energy in bonds or of intermolecular forces

Type of bond or intermolecular force	Example of interacting molecules	Energy in kJ/mol (kcal/mol)
Covalent	$RH_2C\text{–}CH_2R$	400–1200 (100–300)
Ionic	$R_4N^+ \bullet\bullet\bullet {}^-OOC\text{-}R$	200–800 (50–200)
Hydrogen bond	$H_3CO_H \bullet\bullet\bullet HO_H$	20–50 (5–12)
Dipole–dipole	$H_3CC{\equiv}N \bullet\bullet\bullet C_6H_5Cl$	12–40 (3–10)
Dipole-induced dipole	$H_3CC{\equiv}N \bullet\bullet\bullet C_6H_6$	10–25 (2–6)
Dispersion or van der Waals	$C_6H_6 \bullet\bullet\bullet C_6H_{14}$	5–20 (1–5)

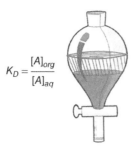

$$K_D = \frac{[A]_{org}}{[A]_{aq}}$$

Figure 2.2 *Distribution of an analyte A between an upper organic phase and a lower aqueous phase*

of these variables. The equilibrium distribution for a substance A is given by the *partition ratio*, which is also called the *distribution constant*:

$$K_D = \frac{[A]_{org}}{[A]_{aq}} \tag{2.8}$$

where $[A]_{org}$ is the concentration of compound A in the organic phase and $[A]_{aq}$ is the concentration of compound A in the water phase.

The distribution constant is a constant relating to a specific molecular species, but often the molecules of a compound can be present as different species, for example by dissociation in the aqueous phase:

$$HA + H_2O \leftrightarrow A^- + H_3O^+ \tag{2.9}$$

or by dimerization in the organic phase:

$$2HA \leftrightarrow (HA)_2 \tag{2.10}$$

These equilibria are normally very fast, and it is therefore appropriate to look at the total distribution of all the species of a compound between the two phases:

$$D_C = \frac{[HA]_{org} + [A^-]_{org} + [(HA)_2]_{org}}{[HA]_{aq} + [A^-]_{aq} + [(HA)_2]_{aq}} = \frac{[HA_{total}]_{org}}{[HA_{total}]_{aq}} \tag{2.11}$$

The *concentration distribution ratio*, D_C, between the two phases can also be converted to the *mass distribution ratio*, D_m, by multiplying the concentrations with the matching phase volumes:

$$D_m = \frac{[HA_{total}]_{org} \cdot V_{org}}{[HA_{total}]_{aq} \cdot V_{aq}} = \frac{((amount\ of\ HA)_{total})_{org}}{((amount\ of\ HA)_{total})_{aq}} \approx \frac{((amount\ of\ HA)_{total})_{stat}}{((amount\ of\ HA)_{total})_{mob}} \tag{2.12}$$

where V_{org} and V_{aq} refer to the volumes of the organic and water phases, respectively. The terms *stat* and *mob* refer, respectively, to the stationary phase and the mobile phase used in chromatography, and this is discussed in more detail in Chapter 4.

The fundamentals of partition are also further outlined in Chapter 6 in relation to sample preparation. The greater the partition coefficient, the higher the affinity toward an organic phase will be. In case of distribution to a solid phase, as in solid phase extraction, the partition can be governed by characteristics other than partition ratios. The partitioning of analytes in a system where one phase is a gas, as in gas chromatography, necessitates that the analytes can enter the gas phase. Discussions on the extraction and partition of compounds therefore most often refer to liquid–liquid systems. Partition ratios are estimated using distribution between n-octanol and water. If the compound can be ionized, the ionized form will have a much stronger affinity toward the aqueous phase as water molecules will solvate the ions. The distribution of an ionizable compound will therefore very much depend on the pH of the aqueous phase. From the Henderson–Hasselbalch equation given above, the following equations can be derived:

$$\text{For acids}: \quad D_{app} = \frac{D_C}{1 + 10^{pH - pKa}} \tag{2.13}$$

$$\text{For bases}: \quad D_{app} = \frac{D_C}{1 + 10^{pKa - pH}} \tag{2.14}$$

If the distribution ratio, D_C, and the pK_a value are known for a compound, the *apparent distribution ratio*, D_{app}, at a given pH can be calculated.

Parameters such as the partition ratios in octanol–water are available as the so-called *log P values*, and the distribution ratio of compounds between octanol and water at different pH values in the water phase is tabulated as *log D values*. Computer programs can also be used for estimation of pKa values, log P values, log D values, and the water solubility of compounds (see Figure 2.3). The actual values of each parameter can vary when consulting different literature references, and this is most often due to differences in the methods used for analysis. This is particularly true for log P, log D, and water solubility data. In a bioanalytical chemical context, such parameters should primarily be used as a guide.

Liquid–liquid extraction (LLE) is often used in sample preparation, as discussed in more detail in Chapter 6. In LLE, it is of interest to determine the fraction of analytes extracted under given conditions. This is given by the general formula:

$$E_n = 1 - \left[\frac{1}{1 + D_c \left(\frac{V_2}{V_1} \right)} \right]^n \tag{2.15}$$

where E_n is the *extracted fraction*, D_c is the distribution ratio between the two phases V_2 and V_1, and n is the number of extractions. V_1 is the phase that originally contains the

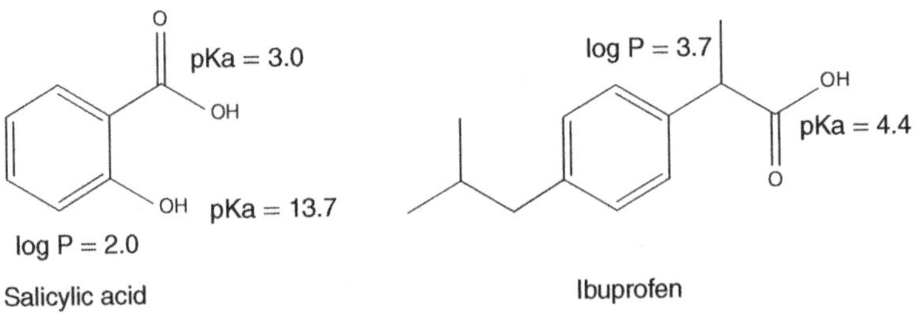

pKa = 3.0

OH

OH   pKa = 13.7

log P = 2.0

Salicylic acid

log P = 3.7

OH

pKa = 4.4

Ibuprofen

Figure 2.3   *Chemical structures of ibuprofen and salicylic acid with log P and pKa values*

analyte, and V_2 is the phase to which the analyte is extracted. In bioanalysis, a high *recovery* is desired in order to minimize the loss of analyte and to improve the reliability of data obtained. Thus, a log P of at least 2 and extraction performed at favorable pH conditions are required. This is illustrated in Boxes 2.2 and 2.3.

Box 2.2 Extraction of Ibuprofen

Ibuprofen has a log P value of 3.72 and a pK_a value of 4.43. If 10 ml of a sample solution of ibuprofen at pH 6.0 is to be extracted to 30 ml of an organic solution, how much will be extracted?

A log P value of 3.72 corresponds to a K_D of 5248. At pH 6.0, the apparent distribution ratio will be $5248 / 1 + 37 = 138$ (using Equation 2.13). Calculating the extracted fraction gives 0.9997, or 99.97%. Doing the same extraction at pH 7.0 will result in an extraction of 97.7%. Try to do this calculation yourself.

Box 2.3 Extraction of Salicylic Acid

Salicylic acid has a log P of 2.0 (corresponding to a K_D of 100) and pKa values of 3.0 and 13.7. Performing similar calculations as in the above example, it can be shown that only 19% of the salicylic acid is extracted into 30 ml of organic phase at pH 6.0. Lowering the pH to 5.0 will give an apparent D_c of about 1 and thus an extraction of 75%.

A question could be if multiple extractions using the same total amount of organic solvent would improve the extraction yield. Consider using three extractions of only 10 ml each of organic solvent. Calculations using Equation 2.15 show that the total extraction in the combined 30 ml will be 87.5% compared to the 75% obtained in only one single 30 ml extraction. Multiple extractions are more efficient, but in the case of salicylic acid it is necessary to perform the extraction at a lower pH value if quantitative extraction is needed.

Similar extraction calculations can be performed for bases using Equations 2.14 and 2.15. It is obvious that quantitative extractions from an aqueous solution into an organic phase are more easily achieved if extraction is performed when the analytes are not ionized. Thus, extraction of carboxylic acid should take place at low pH ($pK_a - 2$–3 pH units) and extraction of bases at high pH ($pK_a + 2$–3 pH units) in the aqueous phase.

2.4   Stereochemistry

Many pharmacologically active molecules, including drug substances, are *chiral*. Biological systems are built from chiral building blocks, proteins and peptides are built from chiral amino acids, polysaccharides are built from chiral monosaccharides, and DNA is chiral due to its α-helix structure. The biological system including the human body is therefore able to distinguish between stereoisomers that differ only in their spatial configuration. This may result in very different pharmacologically responses and different ADME (adsorption, distribution, metabolism, and excretion) properties of such isomers.

To describe the stereochemistry of a drug substance is to visualize the spatial orientation of its components in space. Biological systems contain large biomolecules that are constructed from building blocks with unique stereochemistry.

Figure 2.4 shows how isomers can be divided into several groups. *Constitutional isomers* are, of course, different compounds with different chemical structures. The *diastereomers*, grouped under the stereoisomers, contain compounds in which the isomers have different physicochemical characteristics with different melting points, partition ratios, and so on. These isomers are therefore easy to separate in chromatographic systems. *Cis-trans isomers* belong to this group, and a number of drug substances can be found in this group. Two examples are given in Figure 2.5. However, *enantiomers*, which also belong to this group of isomers, require chiral systems for their separation due to their identical physicochemical properties.

Isomerization at the double bond is often mediated by light, and therefore the compound should be protected from light exposure.

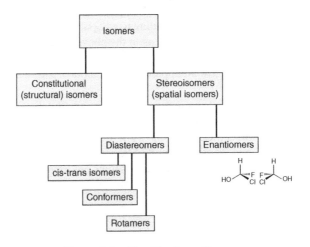

Figure 2.4   *Classification of isomers*

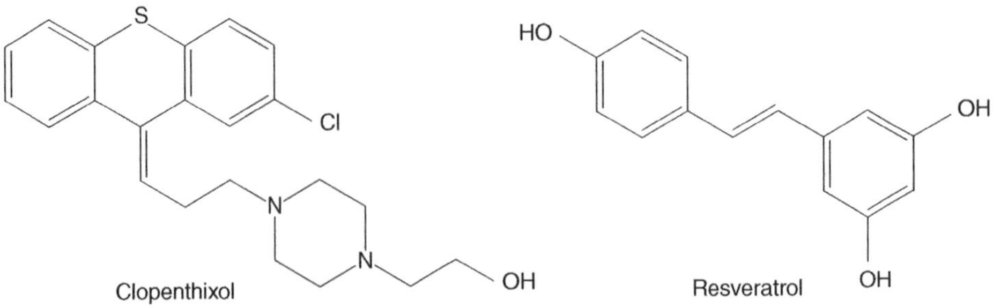

Figure 2.5   *The chemical structure of cis-clopenthixol and trans-resveratrol*

Enantiomers constitute a special group of stereoisomers. The two enantiomers that constitute a pair contain a *chiral center* and are mirror images of each other. A chiral center is an atom connected to three (S and P atoms) or four (N and C atoms) different ligands. The most abundant chiral center is where a carbon atom is connected to four different groups, but nitrogen, phosphor, and sulfur also can be chiral centers. A compound containing one or more chiral centers is able to rotate plane-polarized light either left or right. This is denoted by (−) or (+), respectively. However, this is not an unambiguous way to describe the configuration of the chiral center as the direction and size of the rotation are dependent on the solvent used for sample preparation. In older literature, the terms *d* (dexter) and *l* (laevo) were used to denote (+) and (−), respectively, but also the capital letters *D* and *L* have been used where reference was made to the configuration of glyceraldehyde. To give an unambiguous description of the configuration of the chiral center, the *R/S nomenclature* has to be used. This nomenclature gives the absolute configuration of the position of groups connected to the chiral atom, and this nomenclature should always be used.

If a *racemic mixture* (equal amounts of two enantiomers; Figure 2.6) is given to a human, it is necessary to determine the ADME properties of both enantiomers because these may be different. When racemic ibuprofen is given, the inactive R-ibuprofen will in vivo be converted to the active S-ibuprofen. When the active S-ibuprofen is given, no R-ibuprofen is formed.

In guinea pigs, it has been shown that the S-cetirizine has a higher distribution to brain tissue than the R form due to a difference in plasma binding. In humans, the S-cetirizine has a higher excretion rate than the R form. Thus, it is important to be able to analyze the enantiomers separately. An example of this is described in more detail in Chapter 9.2. Many

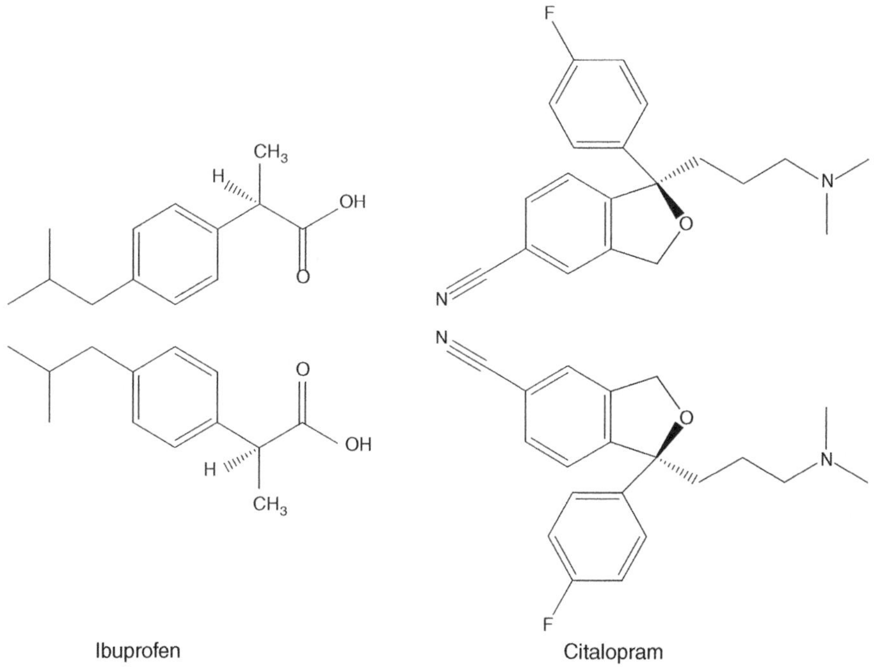

Ibuprofen   Citalopram

Figure 2.6 *Enantiomeric drug compounds*

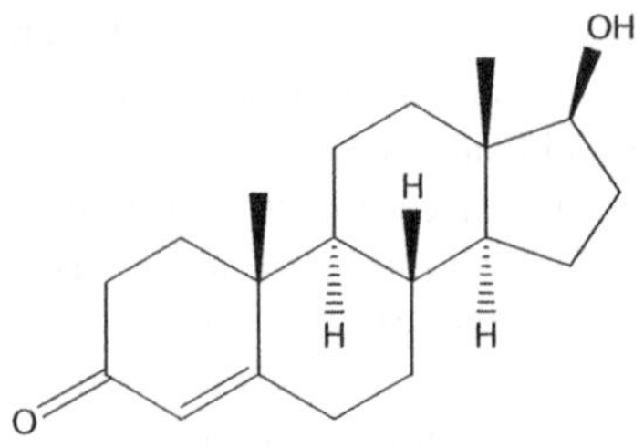

Steroid (testosterone)

Figure 2.7 *A drug substance with several chiral centers*

natural products and endogenous molecules contain several chiral centers (Figure 2.7). These molecules can be considered unique and will not have an enantiomer counterpart.

2.5 Peptides and Proteins

There are many peptide and protein *biopharmaceuticals*. Interferon α2, erythropoietin (EPO), insulin, and human chorionic gonadotropin (hCG) are only a few examples of biopharmaceuticals in use. These compounds are either recombinantly produced or extracted and purified from animal or human sources. Because these are proteins, they also have the same structural features as endogenous proteins. Depending on their structure, situation, and environment, proteins play vital roles in transport, signaling, immunity, and reaction catalysis. Although the diversity in protein structure and function is overwhelming, the basic structure elements are of less complexity.

2.5.1 Building Blocks: Amino Acids

Peptides and proteins are built out of *amino acids*. These are organic compounds consisting of an amino group (NH_2), a carboxylic acid group (COOH), and a side chain (R). Due to the presence of these functional groups, amino acids are zwitterionic. This means that at certain pH values, they both have a positive and negative charge, but overall they are neutral. Additionally, because the side chain is coupled to the α-carbon (see Figure 2.8), all amino acids (except glycine) are chiral. As building blocks for peptides and proteins, mainly L-amino acids are used. Figure 2.8 shows the general structure of an amino acid, its chiral C-atom, as well as their charge profile in relation to pH.

There are, in total, 20 different naturally existing amino acids present in peptides and proteins. These differences are due to variations in the side chains (R) and can be categorized as polar, nonpolar, basic, acidic, aliphatic, and aromatic. These properties affect retention behavior in chromatographic systems and choice of detection principle: the nonpolar side chain of phenylalanine will cause stronger interaction with reversed-phase sorbents than the polar side chain of asparagine. The aromatic side chain of tryptophan will allow specific detection with an UV detector because one of its UV absorbance optima is 280 nm. Amino acids are designated with either a three-letter abbreviation or a one-letter abbreviation. Table 2.3 shows the chemical structure and the most important physicochemical properties of these 20 amino acids.

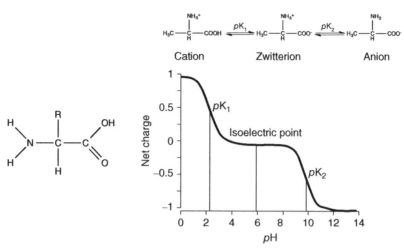

Figure 2.8 *General structure of amino acid, including its chiral C-atom and the charge–pH dependency of amino acids*

Table 2.3 *Amino acid abbreviations and key properties*

Amino acid	Three-letter abbreviation	One-letter abbreviation	Side chain pK_a	Side chain wavelength	Side chain polarity
Alanine	Ala	A	——	——	Nonpolar
Glycine	Gly	G	——	——	Nonpolar
Isoleucine	Ile	I	——	——	Nonpolar
Leucine	Leu	L	——	——	Nonpolar
Proline	Pro	P	——	——	Nonpolar
Valine	Val	V	——	——	Nonpolar
Cysteine	Cys	C	8.3	250	Nonpolar
Methionine	Met	M	——	——	Nonpolar
Phenylalanine	Phe	F	——	257	Nonpolar
Tryptophan	Trp	W	——	280	Nonpolar
Asparagine	Asn	N	——	——	Polar
Glutamine	Gln	Q	——	——	Polar
Serine	Ser	S	——	——	Polar
Threonine	Thr	T	——	——	Polar
Tyrosine	Tyr	Y	10.1	274	Polar
Arginine	Arg	R	12.5	——	Basic polar
Lysine	Lys	K	10.5	——	Basic polar
Histidine	His	H	6.0	211	Basic polar
Aspartic acid	Asp	D	3.6	——	Acidic polar
Glutamic acid	Glu	E	4.2	——	Acidic polar

2.5.2 Composition of Peptides and Proteins

By coupling the amino acids through (bio)synthesis via the so-called *peptide bond*, chains of amino acids are generated. These chains are often referred to as *polypeptide chains*.

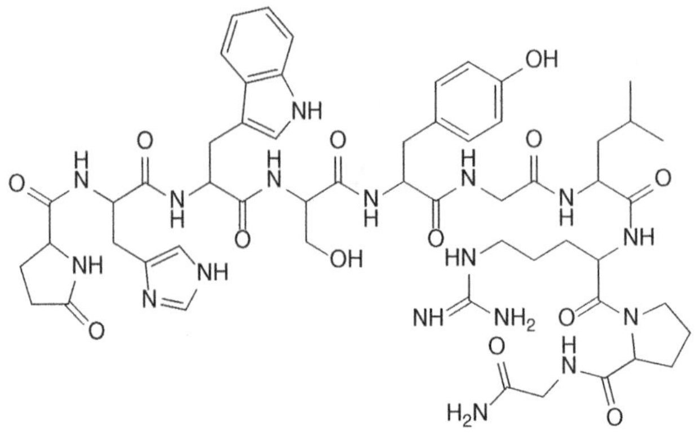

Glp-His-Trp-Ser-Tyr-Leu-Arg-Pro-Gly-NH$_2$

Figure 2.9 *Structure of gonadorelin including three-letter abbreviation*

Within but also between these chains, cysteine residues might through oxidation of their sulfhydryl functional groups form *disulfide bonds*. The chains of amino acids are called either *peptides* or *proteins*, depending on the size of the biomolecule. The size ranges from approximately 150 Da to several hundreds of thousands of Daltons. There is no clear size boundary to be used to differentiate between peptides and proteins. Some sources use 50 amino acids as the limit, whereas other sources state that 10 kDa is the division line. Examples of peptide drugs are the decapeptide gonadorelin (see Figure 2.9) and glucagon (consisting of 29 amino acids). A typical protein drug is EPO.

The larger the chain of amino acids gets, the more flexible the biomolecule gets, resulting in many variations in 3D structure and the formation of complexes of several protein subunits. The structure of a protein can be described on four different levels: primary structure, secondary structure, tertiary structure, and quaternary structure (see Figure 2.10).

The *primary structure* is the sequence of the covalently bonded amino acid chain. It is nothing more than a description of how the amino acids are arranged and which of the cysteine residues have formed disulfide bridges.

The *secondary structure* is due to interactions between closely situated functional groups present in the side chains, and repeating structures occur. These structures are designated as α-helixes, β-sheets, and turns.

The *tertiary structure* is caused by interactions between parts of the protein that are not situated closely together. It describes the shape of the protein and how the secondary structure elements are folded together.

The *quaternary structure* is the combination of two or more polypeptide chains into one complete unit. Each polypeptide chain involved in such a conformation is called a subunit.

2.5.3   Physicochemical Properties of Polypeptides

The physicochemical properties of peptides compared to proteins are very much the same: a peptide is in its composition no more than a very small protein. Sequence elements in the

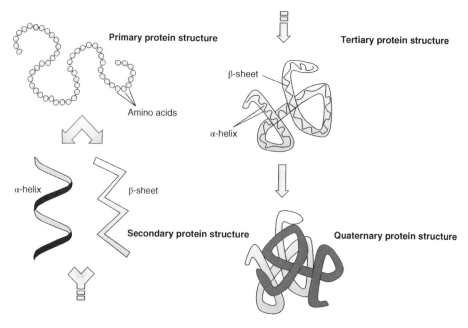

Figure 2.10   *Overview of the different protein structure levels*

primary structure but also through folding in the tertiary structure make parts of a polypeptide nonpolar and other parts polar. Nonpolar areas are mainly situated in the hydrophobic cores of a polypeptide or in more linear parts that are in membranes. In aqueous solutions, the nonpolar regions are usually not exposed. However, in chromatography, conditions can be such (e.g., the presence of reversed-phase sorbent and organic solvents like acetonitrile) that proteins unfold, exposing the nonpolar regions and giving rise to retention based on hydrophobic interactions.

The polar parts of the polypeptides consist of the uncharged and charged side chains. From Table 2.3, it can be seen that the amino acid side chains histidine, lysine, arginine, aspartic acid, and glutamic acid have acid–base properties and as such can influence the overall charge of the polypeptide. The overall charge is determined from the individual pK_a values of each of the side chains as well as the N- and C-terminal functional groups. The pH at which the net overall charge of the polypeptide is zero is called the *isoelectric point* (pI or IEP). At this point, the polypeptide is least soluble. The pI is polypeptide specific. In analogy with pK_a, it is such that when the pH is increased above its pI, the net charge of the polypeptide is negative. When the pH is below the pI, the net charge is positive. The value of the pI can be calculated from the pK_a values from all the side chains with acid–base properties as well as the pK_a values of the N- and C-terminal groups. The calculation itself will not be discussed in this textbook. However, there are several web-based pI calculators available. EPO contains many functional groups with acid–base properties: 25 basic ones (3 histidine, 13 arginine, 8 lysine, and 1 N-terminal amine) and 20 acidic ones (6 aspartic acid, 13 glutamic acid, and 1 C-terminal carboxyl). For this protein, the pI can be calculated to 8.30.

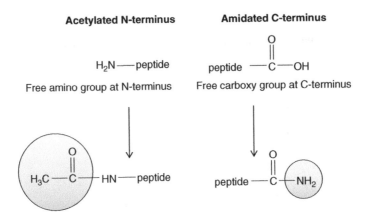

Figure 2.11   *Acetylation and amidation of terminal amino acids in a polypeptide chain*

2.5.4   Peptide Backbone Modifications

Peptides and proteins can be modified at different stages. As early as in the production of biopharmaceuticals, modifications as a result of the (bio)synthesis process might occur. Also, after administration, the biopharmaceutical is prone to modification. All these modifications affect the sample preparation and analysis of the biomolecule.

Production-based modifications occur during synthesis of the peptide or protein. For example, they are often chosen to protect the N-terminal group, C-terminal group, or side chain during resin-based synthesis of a (poly)peptide. These protective groups change both the peptide's polarity and the acid–base properties. An example of N-terminal protection is the use of t-Boc (tert-butyloxycarbonyl protecting group). In the case that such a protecting group is not removed from the polypeptide, there will be no charge on the N-terminal nitrogen, resulting in a change in charge and polarity of the peptide at a certain pH. Another example is the biosynthesis of darbepoietin (also known as novel erythropoiesis-stimulating protein, or NESP), which is produced by recombinant DNA technology in a cell culture. This biopharmaceutical differs not in amino acid sequence but in glycosylation pattern from the endogenous EPO.

During chemical synthesis of biopharmaceutical polypeptides, the N-terminal and C-terminal groups can be different from those occurring naturally. Most seen examples are the acetylation of the N-terminal amino group and/or the amidation of the C-terminal carboxylic acid group (see Figure 2.11). Both modifications lead to changes in net charge of the polypeptide.

Posttranslational modifications (PTMs) occur after administration of the protein biopharmaceuticals. PTMs are endogenous processes leading to a change in structure. These modifications affect the physicochemical properties. An example of PTM is the phosphorylation of the amino acid serine. This leads to an increased polarity and the introduction of a negatively charged group, thus changing the pI. Another example is N-glycosylation of asparagine. Although the pI is not influenced, the polarity increases due to the coupling of a polar sugar moiety to the polypeptide.

3

Biological Samples: Their Composition and Properties, and Their Collection and Storage

Steen Honoré Hansen

School of Pharmaceutical Sciences, University of Copenhagen, Denmark

The test materials to be investigated using bioanalytical methods are very diverse in their nature. They span from exhaled air and biofluids like plasma and urine to breast milk and tissue from biopsies as well as hair and nails. Plasma contains a lot of proteins, whereas urine is 98% water and breast milk contains a lot of fat. Therefore, an individual method with especially dedicated sample preparation is needed for the specific test material. Knowledge of the composition and of other properties of the test material will make it easier to develop suitable sample preparation and measuring techniques. In order to obtain reliable data, it is also important to know how to handle and store the samples.

3.1 Introduction

Biological samples have to be considered as a health risk factor because of the possibility that they are contagious, and precautions should therefore be taken accordingly. The laboratory staff should always use gloves, equipment intended for repeatedly utilization should be disinfected after use, and all other materials should be disposed as biological waste.

Blood and other body fluids as well as tissue samples are biological in nature, and they are rapidly decomposed by microorganisms (bacteria and mold). Therefore, it is absolutely essential that such samples are handled properly during collection and storage. One very important issue in the collection and storage of biosamples is to use labels and text that are waterproof. Otherwise, the information on the vial stored in a freezer may be lost.

Bioanalysis of Pharmaceuticals: Sample Preparation, Separation Techniques and Mass Spectrometry,
First Edition. Steen Honoré Hansen and Stig Pedersen-Bjergaard.
© 2015 John Wiley & Sons, Ltd. Published 2015 by John Wiley & Sons, Ltd.

The best type of sample to use in a given situation depends on the question to be answered. A drug substance may have a different distribution to the red blood cells, to plasma, and to saliva, and analysis of whole blood, plasma, and saliva for such a drug substance will thus result in different data. There will also be a lag time from the dosing of a drug until it can be detected in urine, and this depends on the elimination rate through the kidneys. Whereas data from a plasma sample are a picture of the status at a given time point, data from a urine sample comprise a mean of a certain time span due to the collection of the urine in the bladder before the urine was passed. Analysis of hair and nails has become popular when a retrospective evaluation of a person with respect to their exposure to a drug or chemical is necessary. However, such an investigation cannot start until a couple of weeks have elapsed because the hair or nails have to grow to a certain length before sampling is possible. In this chapter, the composition, other properties, collection, and storage of a number of biomaterials will be presented.

3.2   Blood, or Whole Blood

Whole blood from humans is an inhomogeneous suspension of erythrocytes, leucocytes, thrombocytes, and other minor cell fractions in a matrix called *plasma*. In volume, the cell fraction in total constitutes about 45% (44% of the 45% is erythrocytes), and the plasma volume about 55% (Figure 3.1). The pH of plasma is about 7.4 and is kept within a very narrow range (7.35–7.43), having a good buffer capacity. When analyzing blood, it is important to be aware that the concentration of analytes may differ from that of the internal volume of the cells to the outside surrounding plasma. Analytical data for a given analyte obtained on plasma, erythrocytes, and lysed blood from the same sample of blood may therefore differ significantly.

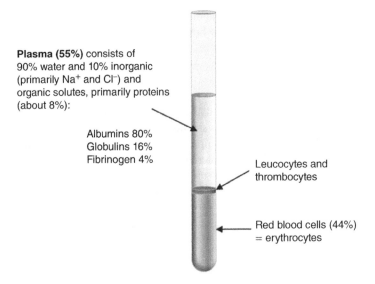

Figure 3.1   *Whole blood after centrifugation*

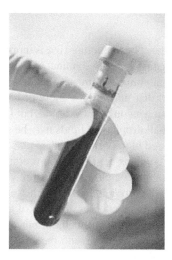

Figure 3.2   *Whole blood sample in its collection vial*

Figure 3.3   *Sampling for dried blood spot*

Whole blood from humans is obtained by vein puncture by a qualified person and is collected in containers (Figure 3.2) with *anticoagulant* added to prevent the blood from coagulation. Many different vials for collection of blood are commercially available, and tubes for different purposes can be found on the Internet.

In samples collected for whole blood analysis, no other additives have been added, and no parts of the blood have been removed. Whole blood is stored at about 4 °C and should not be frozen. A number of anticoagulants can be used, as discussed in Section 3.3. Whole blood samples have to be thoroughly mixed before analysis to counteract any sedimentation of blood cells.

Blood may also be collected in the form of *dried blood spots* (Figure 3.3). A drop of blood is placed on a piece of paper or filter paper, then dried. The sample can then very easily be transported from the site of collection to any laboratory in the world for further analysis. This is an advantage as no storage on dry ice during transport is needed and the transport weight is also considerably less. An example of an analysis using dried blood spots is given in Section 9.3.

3.3   Plasma and Serum

Plasma contains about 90% water. The remaining 10% is inorganic or organic ions (2%) and proteins (about 8%). The inorganic ions are dominated by sodium and chloride, whereas the main part of the proteins is albumin. A minor but not less important part of the proteins is the enzymes (e.g., esterases) that may still be more or less active in collected plasma. Plasma is normally a clear, homogeneous yellow liquid, but if collected from a person who just has eaten a large lunch, small droplets of lipids may be seen on the surface. If plasma is pink or red, *hemolysis* of the red blood cells has taken place. Plasma is stabilized with an anticoagulant, and thus it still contains the fibrinogen proteins that normally will provide coagulation. The anticoagulants may be citrate, EDTA (ethylenediamine tetraacetic acid), or oxalate, which all are good *calcium chelators*. Also, the sulfated polysaccharide *heparin* may be used as an anticoagulant. In most cases, the type of anticoagulant has no influence on the data obtained using the bioanalytical method, but it is of course important to be sure that the anticoagulant does not interfere with the analysis. The use of serum instead of plasma circumvents any problems due to anticoagulants. If small amounts of clots are seen in plasma, it is advisable to centrifuge it before using it for analysis.

Plasma is obtained from whole blood collected in vials containing anticoagulant and then centrifuged typically for 10 minutes at 1000 g. After centrifugation, plasma constitutes the upper, yellow part in the centrifuge vial. Plasma can be stored frozen typically at either −20 or −80 °C.

Serum is obtained from whole blood after coagulation and by centrifugation, typically for 10 minutes at 1000 g. The blood is collected in containers without anticoagulant added and left at room temperature for about 30 minutes to coagulate. After centrifugation, typically for 10 minutes at 1000–2000 g, serum constitutes the upper, yellow part in the centrifuge vial. Serum can be stored frozen, typically at either −20 or −80 °C. The constituents in serum are very similar to those of plasma. The only difference is the lack of the fibrinogen proteins, and they make up only a small part of the protein fraction. Serum, like plasma, is a clear, homogenous liquid, and it should contain no clots because these should already have been removed by centrifugation. Plasma as well as serum are very similar in constituents from person to person.

3.4   Urine

Urine from humans is normally a yellow, clear, and sterile liquid. The contents of solutes in urine may vary a lot, depending on the intake of food and drink. The urine is normally more concentrated in the morning but may be diluted due to increased intake of beverages and alcohol. The average volume excreted per day is 1 l but may easily vary between 500 and 2000 ml. The pH of urine is normally about 6 but may vary between 4.6 and 8 depending on food intake. Intake of ammonium chloride will lower pH of the urine. Urine is the primary excretion route for the many metabolic products formed in the body, of which some (e.g., ammonia) have to be eliminated due to their toxicity. Most of the major solutes in urine are anionic (carboxylic acids) or amphoteric (amino acids) in nature, but the urine contains thousands of different solutes in very different concentrations.

Urine is collected in plastic containers of a suitable volume (Figure 3.4). If the urine samples are to be used for recovery studies in which the total volume of urine has to be

 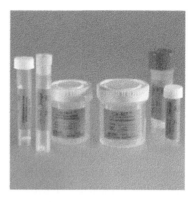

Figure 3.4 *Containers for urine collection*

known, it is important to ask the patient for very careful compliance when collecting. After collection, the volume of urine is measured (and noted), and the major part is discarded. A minor part (5–10 ml) is stored frozen at −20 or −80 °C. When urine is cooled, precipitate may be formed that cannot be re-dissolved when brought back to room temperature. Centrifugation may therefore be needed. It is extremely important to mix thawed urine samples because major fractionation takes place during the freezing process.

Urine from mice and rats is normally collected via the *metabolism cages* in which they are kept. The metabolism cages are built to be able to separate urine and feces when delivered from the animals. Rat urine is much more concentrated compared to urine from humans.

Collection of urine from larger animals like dogs and pigs is a little more difficult as the animals often are allowed to move around in a larger place during experiments. Therefore, it is necessary to insert a catheter, including a bag for the urine collection.

3.5 Feces

Human *feces* are an inhomogeneous, bad-smelling substance that may vary a lot in consistency depending on diet and general wellbeing. The color is due to bilirubin derivatives coming from the degradation of red blood cells, and the smell comes from the microbial activity in the gut. Larger drug molecules above about 500 Da have a high tendency to be excreted through the gall duct and may therefore be found in feces. If a glucuronide metabolite is excreted by the gall, it may be hydrolyzed by bacterial β-glucuronidase in the gut and then undergo reuptake in the body. Bioanalysis of feces is most often performed in relation to recovery studies of drug substances for the diagnosis of digestive disorders. Although being outside the human body, the content of the gut is considered to be the largest "organ" of the human body due to its bacterial activity, which can have an effect on the wellbeing of the person.

Human feces samples are a little problematic to collect due to their large variability in consistency and aroma. Suitable containers have to be used, and it is important to perform some kind of homogenization to be able to redraw a representative sample. The feces samples contain an unknown amount of water in which the water-soluble compound is dissolved, but some compound may also be adsorbed onto the particles in the feces. In

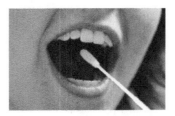

Figure 3.5   *Collection of saliva using a cotton bud*

order to obtain knowledge of the water content, a known representative amount of sample may be dried and then weighed again. Another possibility is to collect the feces in a known amount of a solvent like methanol, homogenize, and then determine the water content of the methanol phase. This will provide knowledge of the total liquid phase, and at the same time the methanol may help to dissolve all compounds for analysis and stop bacterial activity. Collected samples can be stored frozen, typically at either -20 or $-80\,°C$.

3.6 Saliva

Saliva consists of 99.5% water and is excreted from several different glands in the mouth. The saliva can therefore vary in viscosity and pH (from about 6 to 8). It is produced in an amount of about 1 l per day, but in nighttime the production is negligible. The content is inorganic ions and small organic molecules also found in blood, as well as small amounts of proteins and polysaccharides.

Collection of saliva is a *noninvasive technique*, and the collection is rather simple to perform (Figure 3.5). To spit into a collection vial is simple. However, it may be necessary to stimulate the production of saliva by chewing a piece of an indifferent material, for example a piece of PARAFILM™. A number of commercial devices for collecting saliva are available. This can be a tampon of cotton to be placed in the mouth for a few minutes until it is soaked with saliva. Another principle is to use a mouthwash with a solvent in which a dye has been included as an internal standard. This will improve the quantitative data obtained. Collected saliva is stored frozen, typically at either -20 or $-80\,°C$.

3.7 Cerebrospinal Fluid

Cerebrospinal fluid (CSF) is a clear colorless liquid containing a lot of inorganic and organic molecules but in relatively small amounts. The amount of proteins may be up to 1 g per liter but is often less. The fluid is produced in an amount of about 500 ml per day, but most of it is reabsorbed. CSF is collected from the spinal cord and can be used to obtain information on the brain status. CSF can be stored frozen, typically at either -20 or $-80\,°C$.

3.8 Synovial Fluid

Synovial liquid is a clear, colorless, viscous liquid with a protein content (about 2%) between that of CSF and plasma, and furthermore it contains about 3–4 g of hyaluronic acid per liter. The synovial liquid is redrawn from joints (e.g., the knee or elbow) using a syringe and can be stored frozen, typically at either −20 or −80 °C.

3.9 Hair and Nails

The main constituent of hair and nails is *keratin* – a protein with a high content (up to 17%) of the amino acid cysteine, which forms disulfide bridges, making the protein very dense and thorough. The protein fibrillates into fibers. This makes hair and nails difficult to degrade. Furthermore, melanin is also a constituent.

Hair and nails are of special concern as these test materials make it possible to track some history due to storage of analytes in the hair or nails over time. Small amounts of compounds present in blood can be deposited in hair and nails and make these test materials useful for retrospective measurements of special analytes (e.g., narcotics) in samples from a person.

The growth of hair on the head is up to about 1.3 cm per month, whereas hair on other parts of the body grows very slowly. The growth rate of nails very much depends on the age and physical state of the person. Fingernails grow faster than toenails and are said to grow about 3 mm a month. Thus, a full nail has a history of about six months. Samples of hair and nails can be stored at room temperature, but if samples are to be used for determination of a history record, the original orientation of the samples in relation to the test person has to be kept as well.

3.10 Tissue (Biopsies)

Biopsies can be drawn from nearly any tissue (liver, kidney, muscle, brain, colon, prostate, etc.). The biopsies therefore always contain proteins, connecting tissue, and more or less fat. The biopsies are often stored as is or in a *Ringer's solution* at −20 or −80 °C, but storage conditions are very much dependent on the type of analysis that is to be performed.

4

General Chromatographic Theory and Principles

Steen Honoré Hansen

School of Pharmaceutical Sciences, University of Copenhagen, Denmark

The chromatographic techniques involve two phases, of which at least one is mobile and the other most often is stationary. The two phases move relative to each other, and in most cases the mobile phase moves through a bed of a stationary phase. When a mixture of analytes is introduced into these systems, a number of separation mechanisms will influence the partition of analytes between the two phases. In this chapter, an introduction to chromatographic theory and to important separation mechanisms is given. The use of different separation principles will be discussed and illustrated with examples. Special focus is given to liquid chromatography. Gas chromatography will be discussed in Chapter 8.

4.1 General Introduction

The name *chromatography* originates from the basic experiments performed in the very early twentieth century in which colored substances were separated using a system with a polar stationary phase and a nonpolar mobile phase. Analysis of samples containing more than one compound may require some kind of separation to be able to perform an accurate determination of a certain analyte in the mixture. Separation techniques span from simple decantation to sophisticated instrumental techniques based on either rate processes or phase equilibria. Thus, the separation techniques can be classified into three main groups: *particle separation techniques*, techniques based on *phase equilibria*, and techniques based on *rate processes*. The chromatographic techniques belong to the group based on phase equilibria.

The basic principle in chromatography is the partition of analytes between two immiscible phases. When analytes have different partition ratios between the two phases, they

Bioanalysis of Pharmaceuticals: Sample Preparation, Separation Techniques and Mass Spectrometry,
First Edition. Steen Honoré Hansen and Stig Pedersen-Bjergaard.
© 2015 John Wiley & Sons, Ltd. Published 2015 by John Wiley & Sons, Ltd.

can be separated. The distribution ratio of a solute depends on the interaction with the molecules in the two phases. In gas chromatography (GC), few interactions take place between analytes and the *mobile phase*, which is an inert gas because all molecules are far apart in the gas phase. Therefore, the retention and separation in GC are controlled by the choice of *stationary phase* and the instrumental setup (see Chapter 8). In liquid chromatography (LC) covering thin-layer chromatography (TLC) and *high-performance liquid chromatography* (HPLC), interactions of the analyte with other molecules take place in the mobile and the stationary phases, leaving several more parameters to be controlled. In this chapter, the most important separation principles and separation mechanisms used in HPLC will be discussed.

Most bioanalytical methods involving separation techniques take advantage of the versatile HPLC technique. HPLC has become the most used chromatographic technique of all, and this is primarily due to the invention of *reversed phase chromatography* on *chemically bonded phases*. Originally, liquid chromatography was developed using a polar stationary phase and a nonpolar mobile phase. A typical system may consist of polar silica material as the stationary phase and nonpolar organic solvents as the mobile phase (e.g., hexane, heptane, and dichloromethane). Such systems are called *straight phase* or *normal phase* systems as this was the "normal" for many years. The first attempt to reverse the polarity of the chromatographic phases was made around 1942 using particles physically coated with a nonpolar solvent (e.g., liquid paraffin) and using an aqueous mobile phase. With the development of chemically bonded phases in the late 1960s, the reversed phase principle had its breakthrough. Until then, it had been necessary to extract the analytes to an organic phase before performing chromatography, but with the reversed phase systems it became possible to analyze aqueous samples directly. This fact was also of fundamental importance for progress in biotechnology research.

4.2 General Chromatographic Theory

The basic physicochemical principle of chromatography is the partitioning of analytes between two immiscible phases. The two phases move relative to each other, and in most cases one phase moves (the mobile phase) while the other is stationary (the stationary phase). Sample molecules introduced into the chromatographic system will be exposed to one or more interactions (diffusion, collisions, dipole–dipole interactions, hydrogen bonding, electrostatic interactions, etc., as presented in Table 2.2) in the two phases simultaneously with the transport of the analytes through the system. The interactions taking place are dependent on the physical and chemical nature of the analytes as well as on the mobile and stationary phases, and they may result in different partition of analytes between the two phases and thus in different migration in the system. A different migration rate of analytes in the system is synonymous with separation.

In Figure 4.1, a schematic presentation of a chromatographic separation in a column is given step by step. The dissolved sample is initially injected (a) in a given volume onto the column. The different sample molecules will start to travel through the column (b) mediated by the mobile phase (the *eluent*). After a certain time, separation is obtained (c) dependent on the partition of the analytes between the two phases. Observe that the bands of the analytes become broader with the distance traveled and with the time spent in the column. Finally, the analytes are washed out (*eluted*) of the column (d) and detected either

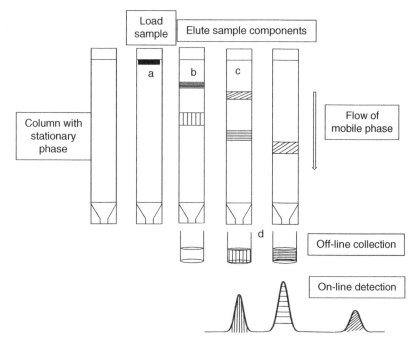

Figure 4.1   *Schematic presentation of a chromatographic separation*

by collection or by an on-line detector. The analytes that have stayed in the column for the longest time also become the broadest band in the resulting chromatogram due to longer exposure to diffusion processes and other interactions in the column.

4.3   Theory of Partition

Differential migration of individual compounds through the column depends on the equilibrium distribution of each compound between the stationary phase and the mobile phase. Therefore, differential migration is determined by those experimental variables that affect this distribution, such as the composition of the mobile phase, the composition of the stationary phase, and the temperature. If we want to alter migration to improve separation, we must change one of these variables. The basic theory of partition is given in Section 2.3, and basic knowledge of the protolytic properties of analytes (which is important for sample preparation as well as for liquid chromatography) is presented in Section 2.2.

The speed by which each compound migrates through the column is determined by the number of molecules of that compound, which is statistically in the mobile phase at any instant, because sample molecules do not move through the column while they are in the stationary phase. The partitioning of molecules between the two phases is, however, a very rapid process. Retention of a compound is therefore determined by its distribution ratio. Compounds with a large distribution ratio statistically have a large portion of their molecules in the stationary phase, and these compounds are strongly retained in the column. Compounds with a small distribution ratio statistically have a small portion of their molecules in the stationary phase, and these are less retained.

A compound appears to migrate smoothly through the column, but at the molecular level the migration of a compound is highly discontinuous. Each time a molecule is distributed to the stationary phase, the migration is temporarily stopped while other molecules of the same kind pass further on. Some of these are immobilized a moment later and will be overtaken by the first molecule, and so on. Each molecule will thus follow a rapid, random alternation between the stationary and the mobile phases in which the concentration of a compound in the stationary phase and in the mobile phase is determined by the partition ratio of the analyte.

For an analyte with protolytic properties, a change in pH will of course have an influence on the observed partition ratio when the degree of ionization is changed. It is important to remember that the protolytic equilibria are much faster than the partition equilibria. Therefore, chromatography cannot be used for separation of the ionized species from the unionized species of an analyte.

4.4   Retention

The *retention volume*, V_R, of an analyte is the volume of mobile phase needed to elute the analyte through the chromatographic system:

$$V_R = t_R \times F \tag{4.1}$$

where t_R is the corresponding *retention time*, and F is the flow rate of the mobile phase in ml/min (Figure 4.2).

The *holdup volume* (also named the *dead volume*), V_M, corresponds to the volume of mobile phase present in the system between the injection point and the detection point.

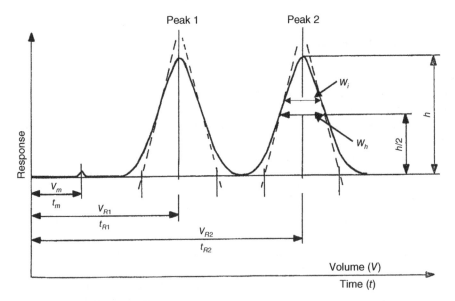

Figure 4.2   *A schematic chromatogram showing relevant parameters*

This volume has to be moved out of the system before an analyte that is injected and only present in the mobile phase can reach the detector.

When an analyte, HA, is eluted to its maximum concentration, half of it has left the system in the mobile phase and the other half is still in the system distributed between the mobile and the stationary phase:

$$V_R \cdot [HA]_M = V_M \cdot [HA]_M + V_S \cdot [HA]_S \qquad (4.2)$$

where V_S is the volume of the stationary phase and $[HA]_M$ and $[HA]_S$ are the concentrations of HA in the mobile phase and the stationary phase, respectively.

Thus:

$$V_R = V_M + V_S \frac{[HA]_S}{[HA]_M} = V_M + V_S D_C \qquad (4.3)$$

and merging Equations 2.12 and 4.3 leads us to:

$$V_R = V_M + V_M D_m = V_M(D_m + 1) \qquad (4.4)$$

where D_m is the *mass distribution ratio* of the analyte and is often referred to as the *retention factor, k*.

For most purposes, *retention time* is used instead of *retention volume*, resulting in the equivalent formula:

$$t_R = t_M(k + 1) \text{ and } k = \frac{t_R - t_M}{t_M} \qquad (4.5)$$

The correlation between retention and distribution ratio in a chromatographic system is obvious, and the use of k makes it easy to compare chromatographic systems that have the same stationary and mobile phases but different column dimensions and connecting tubes, when used at different laboratories.

4.5 Separation Efficiency

Molecules from a number of compounds in a mixture will, after a chromatographic separation, appear in separate bands if the compounds have different D_m. All molecules will migrate back and forth between the mobile and the stationary phases, and the amount of each type of molecule present in the two phases is a statistical question that depends on the distribution ratio of each compound. The faster this exchange of molecules between the two phases is, the better will differences in partition ratios between analytes be expressed, and the more efficient the separation will be. In principle, the molecules of an analyte should be brought into equilibrium between the two phases to express the distribution ratio. This will theoretically take place a number of times during the migration through the column. Each equilibrium is equivalent to a theoretical plate also called *the height equivalent to a theoretical plate*, H. This theory is similar to the theory of theoretical plates in a distillation column.

The number of *theoretical plates*, N, in a column is equivalent to the length, L, divided by H:

$$N = \frac{L}{H} \qquad (4.6)$$

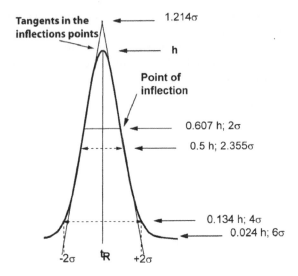

Figure 4.3   *A Gaussian peak with relevant parameters assigned*

The plate height has the length σ_2/x, where σ is the standard deviation of the *Gaussian peak* (see Figure 4.3) being equivalent to one-fourth of the peak width at baseline, w, and x is the distance traveled. For a peak just leaving the column, the following equation is valid:

$$N = \frac{Lx}{\sigma^2} = \frac{16L^2}{w^2} \tag{4.7}$$

And, when using time units for column length and peak width, we get:

$$N = 16\left(\frac{t_R}{w}\right)^2 = 5.54\left(\frac{t_R}{w_h}\right)^2 \tag{4.8}$$

where w_h is the *peak width at half height*. It is therefore easy to calculate the number of theoretical plates for a column. The number is dimensionless and the narrower the single peak, the more peaks can be separated. The number of peaks possible to separate in a given chromatographic system assuming Gaussian peaks and a resolution of 1 is expressed as the *peak capacity*, PC:

$$PC = 1 + \frac{\sqrt{N}}{4}\ln\left(\frac{t_R}{t_M}\right) \tag{4.9}$$

4.6   Resolution

The separation of two closely eluting, Gaussian peaks can be expressed by the *resolution*, R_S:

$$R_S = 2\left(\frac{t_{R2} - t_{R1}}{w_1 + w_2}\right) = 1.18\left(\frac{t_{R2} - t_{R1}}{w_{h1} + w_{h2}}\right) \tag{4.10}$$

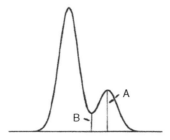

Figure 4.4   *Peak-to-valley ratio between two not fully resolved peaks*

where t_{R2} and t_{R1} are the retention of the last and the first elution peak, respectively. The peak width may be measured at baseline, w, or at half width, w_h. Chromatographic computer software normally uses measurements at half height in their calculations.

When a resolution (Rs) of 1.0 is achieved, an overlap between the peaks is about 3%; and when Rs = 1.5, an overlap of only 0.2% is present. Thus, with Rs ≥ 1.5, the peaks are considered to be fully separated. However, this is only true when the peaks in question are of similar heights. If large differences in peak heights are present, the minor peak may be compromised by a larger overlap from the major peak. In such cases, another parameter for the resolution based on a *peak-to-valley* measurement can be used for characterization of the separation (Figure 4.4), given as the ratio A/B.

An increase in resolution may be achieved in two different ways: (i) by increasing the difference in retention (this is an increase in selectivity, α; see below), or (ii) by achieving narrower peaks (this is an increase in efficiency, N).

4.7   Selectivity

The relative retention of two adjacent peaks in the chromatogram is described by the *separation factor*, α, given by the equation:

$$\alpha = \frac{k_2}{k_1} \tag{4.11}$$

where k_2 is the retention factor of the latter of the two eluting peaks and k_1 is the retention factor of the first eluting peak. The separation factor is a measure of the selectivity of a given chromatographic system. It is a constant for a given set of analytical conditions (mobile and stationary phases) and is independent of the column dimensions. The separation factor is ≥1.0. When $\alpha = 1.0$, separation is not possible. The larger the value of α, the easier a separation becomes. Changes in α are achieved by changing one or more of the parameters: temperature, mobile phase, or stationary phase.

The resolution, R_S, may also be expressed by using the approximate formula:

$$R_S = \frac{1}{4}\sqrt{N}\left(\frac{\alpha - 1}{\alpha}\right)\left(\frac{k}{k+1}\right) \tag{4.12}$$

where efficiency (N), selectivity (α), and retention (k) are represented. From this, it is directly deducible that it requires four times the number of theoretical plates to double

Table 4.1   *The relationship between α and the number of N needed to obtain a resolution of 1.5*

α	$\left(\frac{\alpha}{\alpha-1}\right)^2$	Necessary number of N for $R_S = 1.5$ and $k = 2$
1.01	10 201	826 281
1.02	2 601	210 681
1.03	1 177	95 377
1.04	676	54 756
1.05	441	35 721
1.10	121	9 801
1.15	58	4 418
1.20	36	2 916
1.25	25	2 015
1.30	19	1 514

Table 4.2   *Relationship between the retention factor, k, and the efficiency, N*

Retention factor k	Necessary number of N to obtain $R_S = 1.0$ for given α	
	α = 1.05	α = 1.10
0.1	853 780	234 260
0.2	254 020	69 700
0.5	63 500	17 420
1.0	28 220	7 740
2.0	15 880	4 360
5.0	10 160	2 790
10	8 540	2 340
20	7 780	2 130

the resolution, and as N is proportional to the length of the column, an increase in analysis time of 4 is the result. Therefore, a change in the selectivity is to be preferred. If it is possible to increase α from 1.05 to 1.1, the need for theoretical plates is decreased with a factor of 4 (Table 4.1).

From Table 4.2, it can be learned that some retention is needed; otherwise, a too high number of N is required to obtain separation. The retention factor should be between 1 and 5 to get the best performance out of the chromatographic system. Longer retentions do not increase the separation much but result in long retention times.

4.8   The Separation Process

The molecules in a sample are exposed to a number of physical actions on their transport through the chromatographic system. The molecules are introduced in the mobile phase and will interact with the stationary phase during the chromatographic process. One of the important parameters for the result of the separation is the flow rate of the mobile phase. Already in the 1950s, van Deemter and coworkers studied the effect of the mobile phase on the efficiency of a gas chromatographic separation, and they could express the efficiency in the formula:

$$H = A + \frac{B}{u} + Cu \tag{4.13}$$

This formula (the *van Deemter equation*) expresses the correlation between the efficiency (given by H), the height equivalent to a theoretical plate, and the *band-broadening* phenomena A, B, and C as a function of the flow rate, u.

The A term covers two major influences: (i) that the molecules will travel different distances on their way through the chromatographic system (like in an eddy) due to the non-uniform packing of the particles of the stationary phase, and (ii) that the flow rate through channels is different in the middle compared to the flow rate close to the side walls of the channels, where the flow rate approaches zero. Both phenomena will give rise to broadening of the analyte into a larger volume (Figure 4.5).

The A term is eliminated when using open tubular columns as in capillary GC.

The B term takes the diffusion of the molecule (Brownian movements) in the mobile phase into account. The analyte molecules will diffuse in all directions, and the radial diffusion will not affect the efficiency. But the diffusion in the direction of the length of the column, the *longitudinal diffusion*, will result in band broadening, especially if the flow rate is low. The B term becomes insignificant at higher flow rates. The B term is of more importance in GC than in LC, which is due to the much larger diffusion velocities in the gas phase than in the liquid phase.

The C term covers the *mass transfer* between the two phases (Figure 4.6). This involves the direct back-and-forth transfer of molecules between the mobile and the stationary phases, but it also involves the transfer to stagnant pools of the stationary phase in closed pores of the porous particles (Figure 4.7).

The van Deemter equation can also be visualized as shown in Figure 4.8. The van Deemter equation has been worked on by many researchers, and a more detailed equation

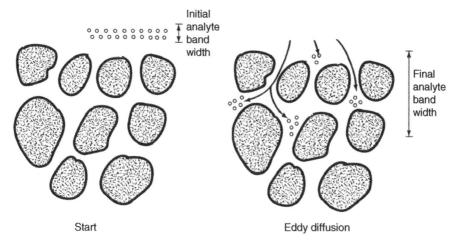

Start   Eddy diffusion

Figure 4.5 *Peak broadening due to eddy diffusion*

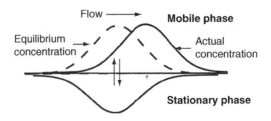

Figure 4.6   *Illustration of mass transfer between the mobile and stationary phases*

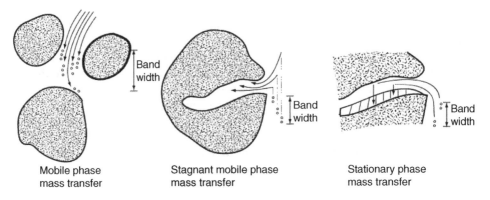

Figure 4.7   *Illustration of mass transfer in the mobile and stationary phases*

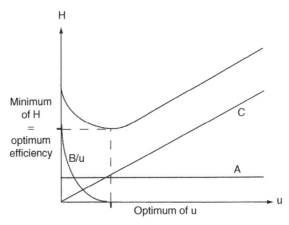

Figure 4.8   *A schematic presentation of the van Deemter plot showing the optimum efficiency (at minimum H) and the optimum flow rate, u*

used for HPLC could, for example, be:

$$H = 2\lambda d_p + \frac{2\gamma D_m}{u} + \frac{\omega(d_p \; or \; d_c)^2 u}{D_m} + \frac{R d_f^2 u}{D_s} \tag{4.14}$$

or

$$H = C_e dp + \frac{C_m dp^2 u}{D_m} + \frac{C_d D_m}{u} + \frac{C_{sm} dp^2 u}{D_m} + \frac{C_s d_f^2 u}{D_s} \qquad (4.15)$$

where D_m and D_s are the diffusion in the mobile and the stationary phases, respectively; d_p is the particle diameter of the particles constituting the chromatographic bed; d_c is the capillary diameter (used for open tube capillary GC or capillary LC); and d_f is the layer thickness of the stationary phase. Λ, λ, γ, ω, R, C_e, C_m, C_d, C_{sm}, and C_s are constants. From this, it is obvious that the particle diameter, dp, is important, and the plate height will decrease when the particle diameter decreases. A smaller plate height will give room for more plates in a given column length, and the efficiency will thus increase. Similarly, a decrease in internal diameter of a GC capillary column will increase the efficiency.

4.9 Chromatographic Principles

4.9.1 Normal Phase Chromatography

Normal phase chromatography, also called straight phase chromatography, is the most common separation principle in TLC but can also be performed in HPLC mode. A polar stationary phase is used along with a more nonpolar mobile phase. A typical choice could be the use of bare silica as the stationary phase and a heptane–ethyl acetate or heptane–propanol mixture as the mobile phase. Both polar and nonpolar small molecules can be separated using this type of chromatography, but it is not suited for larger molecules like proteins because of the high content of organic solvent in the mobile phase. When analytes enter such a system, the analytes will interact with the stationary and the mobile phases. Analytes with no affinity to (no interaction with) the polar stationary silica will not be retained and will travel with the speed of the mobile phase. Analytes having polar functional groups will have a higher affinity to the polar silica and will show retention in the system. The adsorption onto the stationary silica is a reversible interaction, and an increase in the polar component of the mobile phase will increase the competition between analyte molecules and polar molecules from the mobile phase for the adsorption sites on the surface, thus weakening the interaction of the analyte with the silica. When the elution strength of the mobile phase in this way is increased by increasing the polarity, the retention of analytes will decrease.

4.9.2 Silica

Bare *silica*, or silica gel, is the most common stationary phase used for normal phase chromatography. Silica has strong adsorption characteristics and is among other things used as a desiccant, and many other substances than water can be adsorbed onto silica. Normal phase chromatography is also called adsorption chromatography. Silica is a partially dehydrated form of colloidal polymerized silicic acid. Silicic acid, H_2SiO_4, does not exist as a free monomer but is available in the form of a sodium silicate solution. When the sodium silicate solution is acidified, the polymeric silica is formed. Silica used for chromatography undergoes an extensive purification process to remove metal impurities and is then pulverized, dried, and fractionated into appropriate particle sizes. However, today most silica materials are made in a spherical form and with a very narrow particle size distribution using organic alkoxysilane compounds as starting materials. Silica has typically a large surface area in the range from 200 to 800 m² per gram. The large surface area is due to the structure of the silica being a fully porous material similar to a sponge. The pore volume of 1 g is more than 0.7 ml. In practice, this means that silica is a porous skeleton that is manufactured with a well-defined pore diameter, typically in the range from 60 to 150 Å (1 Å is 10^{-10} m) when used for chromatography of small molecules. When the mobile phase flows through the silica, it will enter the entire volume between the particles and the whole pore volume inside the particles. This provides a tremendous network of contacts between the stationary and the mobile phases, and analytes will be exposed to this surface area of the stationary phase.

The silica is covered with *silanol groups* on the surface (−Si−OH), which are the adsorptive groups. The silanol groups make the surface polar and behave as weak acids. Figure 4.9 shows a schematic drawing of silanol groups on the surface of silica.

Some silanol groups provide hydrogen bonding interactions with nearby silanol groups, whereas other silanols are isolated. In some cases, two silanol groups are attached to the same silicium atom. The structure differences between silanol groups mean that they have different activity. If a silanol group forms a hydrogen bond to a neighboring group, it will be less active in adsorption processes than an isolated one. Performing chromatography in a highly activated normal phase system where the silica is dry and the organic mobile phase is without water can be difficult to control as the most active silanol groups readily adsorb polar molecules. If small amounts of water are introduced into the system, the water will be adsorbed and will gradually mask the active sites and in this way gradually generate a stationary phase of water. Thus, it is a good idea to include small amounts of water or other polar solvent into the mobile phase to bring it in equilibrium with the stationary phase.

Some of the silanol groups will be negatively charged at neutral pH and will participate in ionic interactions with positively charged analytes. Ionic interactions are often much stronger than other polar interactions. It can therefore be difficult for solvents to displace the analytes from the strong adsorption sites at the stationary phase, resulting in peak tailing of the analyte. The problem can be overcome by adding an amine like diethylamine to

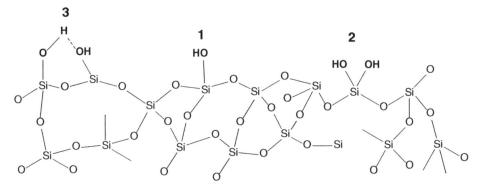

Figure 4.9   *Silanol groups: (1) free (isolated) silanol, (2) germinal silanols, and (3) associated silanols on the surface of silica*

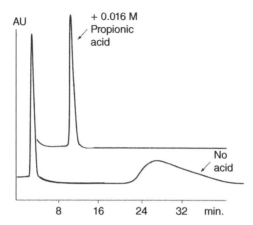

Figure 4.10 *Chromatogram of hydroxyatrazin before (showing strong tailing) and after addition of a carboxylic acid to the mobile phase. HPLC system: column silica 120 mm × 4.6 mm, 5 μm, with dichloromethane + methanol (95:5 v/v) as mobile phase without and with propionic acid added*

the mobile phase. The amine will then compete with the analytes for the adsorption sites and will displace the analyte, forming a more Gaussian peak. Another way to overcome peak tailing of basic analytes is to add a carboxylic acid to the mobile phase. The chromatography is now performed at a lower pH, which suppresses the ionization of the silanol groups, reducing ionic interactions. In Figure 4.10, an example of this is given.

Addition of an amine or an acid to the mobile phase has the same effect and will also help to control the ionization of the analytes. The silanol groups are weakly acidic, and the differences in their structure also result in differences in acidity. The pK_a value of pure silica is therefore an average value of all the silanols and is in the interval from 6 to 7. Silica can therefore also act as a cation exchanger. Furthermore, the silica has an isoelectric point at a pH of about 2 because of increasing surface protonation of the polymer at low pH.

The traditional form of normal phase chromatography using non-aqueous mobile phases is not in general suitable for bioanalysis because small polar molecules (e.g., water) as well as polar bio-macromolecules tend to stick to the silica surface.

4.9.3   HILIC

A more versatile form of normal phase chromatography has become popular in bioanalysis over the past 20 years as it is compatible with mass spectrometry (MS) detection. The principle is called *hydrophilic interaction liquid chromatography* (HILIC). It is in fact possible to use silica as the stationary phase together with mobile phases of very different polarity—from aqueous buffer to the nonpolar heptane.

In Figure 4.11, the separation of five opium alkaloids is shown, and the separation mechanism using 90% water in the mobile phase can be considered cation exchange. With increasing amounts of organic solvent in the mobile phase, the retention decreases, as expected. But at a high content of organic phase, the retention of the polar analytes increases due to a normal phase mechanism where the most polar analytes elute late.

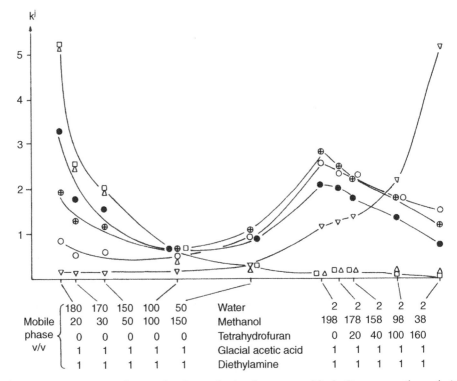

Figure 4.11   *Retention factor of opiates obtained on unmodified silica versus the polarity of the mobile phase. Opiates: +, codeine; ○, morphine; v, normorphine; □, noscapine; Δ, papaverine; and •, thebaine*

HILIC is a chromatographic principle used for the separation of relatively polar analytes where the reversed phase mode does not provide sufficient retention. The mobile phases used in HILIC often have a content of a relatively polar organic solvent (e.g., methanol or acetonitrile) between 60 and 95%. The stationary phases are very polar (e.g., silica or silica derivatized with polar groups (diols, twitter ions, etc.)). The systems can therefore be considered as highly deactivated normal phase chromatography systems, where a hydrophilic stationary phase is present and where ionic interactions also are possible. The less water is present in the mobile phase, the stronger the retention of the analytes. Thus, water is the strongest solvent in HILIC. If gradient elution is applied in HILIC, it is initiated using a high content of organic solvent and progressing with increasing amounts of water.

The HILIC mode has become very popular for analysis of polar and ionic substances in situations where reversed phase chromatography has been less successful. The content of organic solvent is typically above 70%, and this will ease the spray in electrospray MS and thus improve sensitivity in liquid chromatography–mass spectrometry (LC-MS) methods. Although HILIC is a normal phase mode, the content of water in the mobile phase still makes it possible to use this mode for bioanalysis, which is more problematic when using non-aqueous normal phase chromatography. The initial mobile phase when using gradient elution in this mode typically contains 90–95% organic solvent, and during the gradient an increasing amount of water is added. It is therefore important to remember the rule that the sample should be dissolved in the mobile phase or something that is less strongly eluting. In this case, the organic solvent is the less strongly eluting solvent, whereas water is the stronger solvent. Injection of samples with a high content of water may therefore ruin the separation.

The interactions between polar groups like silanol groups on silica and functional groups on the substances to be separated are called polar interactions and are primarily driven by dipole–dipole interactions, hydrogen bonding interactions, and ionic interactions. Silica can act as a cation exchanger and therefore ionic interaction is important when analyzing basic analytes. Such interactions are normally undesirable because they often cause tailing during separation. Ionic interactions can be very strong and comparable to covalent bonds (see Table 2.2) but can be avoided by either masking the most acidic silanol groups or selecting mobile phases that prevent the strong ionic interactions. It is very common to use trifluoroacetic acid or other carboxylic acid as an additive to the mobile phase in order to keep the silanols on a non-ionized form.

4.9.4   Other Stationary Phases for Normal Phase Chromatography

Alumina, zirconia, and magnesium silicates are examples of other polar materials to be used as stationary phases in normal phase LC. The silanol groups on silica can also be derivatized with ligands containing other polar groups such as diol, CN, and NH_2 (chemically bonded phases). These polar materials can provide changes in selectivity and thus changes in the order of elution compared to those obtained on silica.

4.10   Reversed Phase Chromatography

Reversed phase chromatography is the most important separation principle in liquid chromatography. In reversed phase chromatography, the stationary phase is hydrophobic, and the mobile phase is a more polar aqueous solution. This type of chromatography is generally applicable to small molecules, peptides, and to some degree also proteins. However, very polar molecules can be difficult to retain on the hydrophobic stationary phases. In this section, focus is on different stationary phases, mobile phases, and the separation principle used in reversed phase chromatography.

4.10.1   Stationary Phases

Column packing materials used as stationary phases for reversed phase chromatography are typically made of silica derivatized with reagents to form a more or less hydrophobic surface on the silica particle. It is typically obtained by binding hydrophobic groups to the silanol groups using chlorosilanes or other organic silane reagents.

By varying the substituents R' and R, stationary phases with different properties are obtained. The examples of stationary phases shown in Figure 4.12 are ranked after declining hydrophobicity. Octadecyl (C18) is the most hydrophobic phase, and cyanopropyl (CN) is the least hydrophobic of these phases. By far, most of the chromatography performed in reversed phase mode makes use of the C18 materials, which are also known as ODS (octadecylsilane) materials.

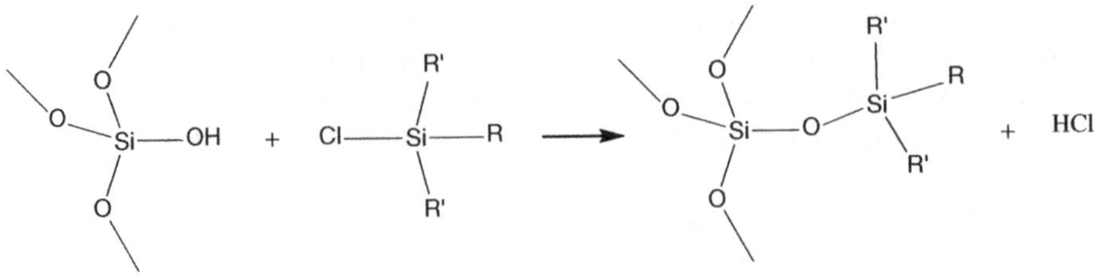

R' = Typically CH$_3$

R = CH$_3$-(CH$_2$)$_{17}$-;Octadecyl
CH$_3$-(CH2)$_8$-;Octyl
C$_6$H$_5$-(CH$_2$)$_3$-;Phenyl
CH$_3$-(CH$_2$)$_3$-;Butyl
CN-(CH$_2$)$_3$-;Cyanopropyl

Figure 4.12   *Derivatization of silica with a chlorosilane reagent*

The surface of silica is covered with silanol groups placed fairly close together. The derivatizing silane reagent has three alkyl groups connected to the silicium atom and is therefore a bulky group. Due to steric hindrance, it is not possible to derivatize all silanol groups with the reagent. A significant percentage of the silanol group can therefore still be present after derivatization, and in order to minimize the number of free silanol groups the material can then be treated with trimethylchlorosilane. This process is named *end capping*. However, even after end capping, some free silanol groups will remain on the surface and be accessible for polar interactions (Figure 4.13), for example ionic interactions with amines.

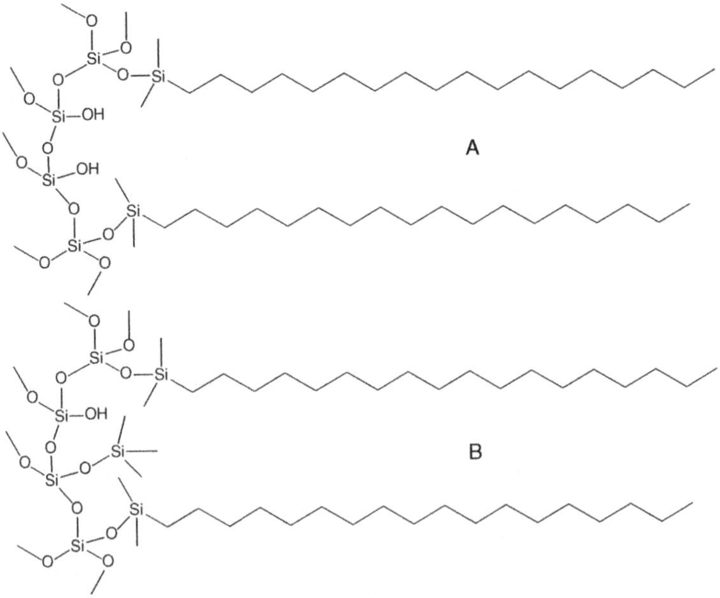

Figure 4.13   *C18 column packing material (A) before and (B) after end capping with trimethylchlorosilane*

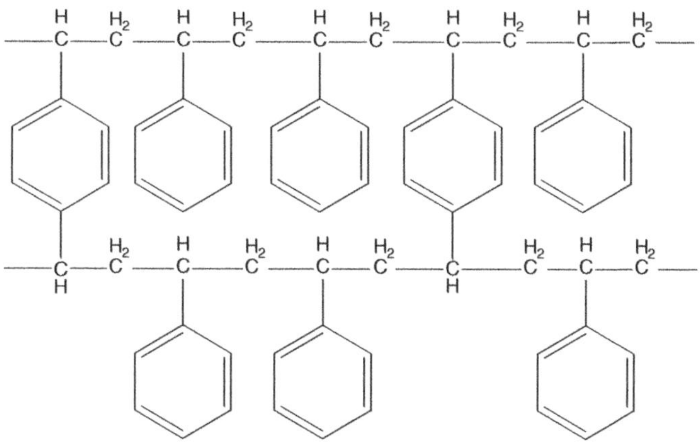

Figure 4.14 *Polystyrene–divinylbenzene copolymer*

Silica-based stationary phases can be used with mobile phases having pH in the range of 2–8. Silica dissolves in an alkaline environment above pH 8 and also to some extent in an acidic environment with a pH lower than 2. It is the *siloxane bonds* that are cleaved. However, it is now possible to obtain silica-based polymers that are stable in the whole pH range by including ethane bridges in the silica polymer. These materials can be derivatized in the same manner as described above. The co-polymerization with organic molecules also has the effect that the density of silanols on the surface decreases, and after derivatization only a few free silanols are left over. When selecting a C18 material for chromatography, it is important to be aware of in which pH interval it can be used. When performing chromatography outside the pH range of 2–8, another possibility is to use pH stable organic polymers like polystyrene-divinylbenzene (PS-DVB). The structure of the PS-DVB copolymer is shown in Figure 4.14.

PS-DVB can be used in the pH range of 1–13, and it often provides a stronger retention of analytes compared to silica-based C18 materials, but the column efficiency of the organic polymeric stationary phases is less than that for silica-based materials. Furthermore, the organic polymeric stationary phases are less rigid than silica, and only highly cross-linked materials are suitable for HPLC.

Activated carbon is an example of another nonpolar adsorbent. Hydrophobic substances in an aqueous environment will adsorb to the hydrophobic surface of the carbon. Recently also diamonds, which are also made of carbon, have been introduced as column packing materials for reversed phase chromatography. These latter materials are, however, not often used in bioanalysis.

4.10.2 Retention Mechanisms

The main separation mechanism in reversed phase chromatography is hydrophobic interactions. Nonpolar analytes are therefore retained strongly, whereas more polar analytes elute earlier. ODS (or C18) column packing materials are the most hydrophobic of the usual commercial stationary phases available. The main forces of interaction are van der Waals forces, which are relatively weak forces but are present in a high number per molecule, and interaction will thus increase with molecular size. The interactions take place between the

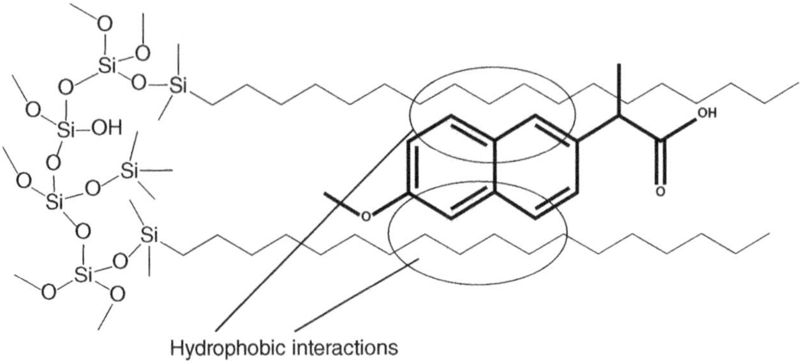

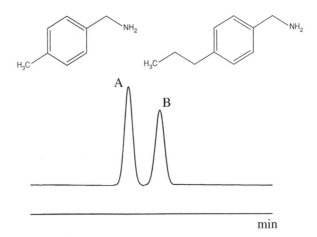

Figure 4.15   *Hydrophobic interactions between the hydrocarbon chain of C18 material and the hydrophobic parts of naproxen*

Figure 4.16   *Chromatogram of two analytes with different sizes of side chain. The separation was obtained using reversed phase chromatography*

hydrophobic hydrocarbon chains of the stationary phase and the hydrophobic parts on the analyte molecule, as shown for naproxen in Figure 4.15.

Even highly polar analytes will have a small retention due to the presence of minor hydrophobic parts in the molecule. Organic analytes having the same functional group will be separated according to the size of the hydrophobic moiety, thus an additional CH_2 group will increase retention. This is illustrated in Figure 4.16. Analyte B is retarded stronger than analyte A because it has a longer hydrophobic side chain.

Besides the hydrophobic interactions with the hydrocarbon chains, a secondary retention mechanism can be displayed by the remaining silanol groups. Polar analytes, especially amines, can have a high affinity for the silanol groups besides the hydrophobic interactions. The resulting separation selectivity is therefore dependent on how many remaining silanol groups are present and how well they are covered.

Table 4.3   *Mobile phases with similar eluting strength*

Mobile phase	A (%)	B (%)	C (%)
Methanol	60	—	—
Acetonitrile	—	46	—
Tetrahydrofuran	—	—	37
Water	40	54	63

For substances with ionizable functional groups, retention will depend on whether these groups are ionized or not. Retention decreases with increasing ionization, and this is of course a function of the change in the mass distribution ratio for such analytes at different pH values.

4.10.3   Mobile Phases

Mobile phases for reversed phase chromatography consist of mixtures of water and one or more organic solvents that must be miscible with water. The organic solvents used are called *organic modifiers* as they modify the eluting strength of the mobile phase. Increased content of organic modifier increases the strength of the mobile phase, and retention of analytes decreases. The solvent strength of methanol is somewhat weaker than that of acetonitrile, which is weaker than that of tetrahydrofuran (THF). The mixtures (A, B, and C) shown in Table 4.3 have about the same solvent strength; they are *isoeluotropic*.

A mobile phase of 60% methanol in water has about the same *eluting strength* as 46% acetonitrile in water or 37% THF in water. When performing a separation of a mixture of compounds using the isoeluotropic phases A–C, the average retention of the compounds will be about the same magnitude for the three mobile phases. However, the order of elution between the substances can be somewhat different because the selectivity in the three systems can be different. Changing the organic modifier can thus be used to change the separation selectivity (Figure 4.17).

Methanol is cheaper and less toxic than acetonitrile. The main drawback when using methanol as a modifier is the formation of more viscous mixtures with water, creating an increase in back pressure in the HPLC system. This is also the reason why ethanol is only seldom used: because the *viscosity* becomes even higher in mixture with water. Acetonitrile may be better suited for UV detection at low wavelength, and it does form less viscous mixtures with water than methanol (Figure 4.18).

Retention of neutral substances is only controlled by the content of organic modifier in the mobile phase and is not affected by pH. For substances with ionizable functional groups, retention is dependent on both the content of organic modifier in the mobile phase and pH.

pH in the mobile phase is controlled by the addition of buffers, and a number of suitable buffers for HPLC/UHPLC (ultrahigh-performance liquid chromatography) are given in Table 4.4. The buffer concentration in the final mobile phase is typically in the range from 0.01 to 0.05 M. Phosphate buffers have for many years been the first choice because phosphate buffers have good buffering properties and low UV absorbance. A problem with phosphate buffers is that they have poor solubility in organic solvents and thus can precipitate at high concentration of organic modifier in the mobile phase. After the introduction

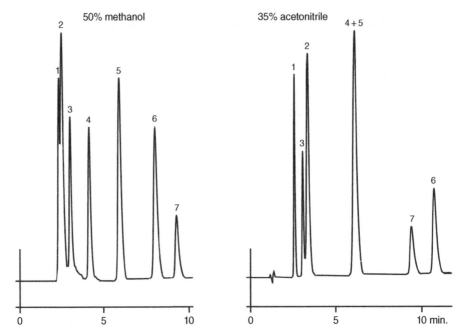

Figure 4.17   *Example of solvent selectivity. Separation of 7 test solutes on a C18 reversed phase HPLC column using (a) methanol or (b) acetonitrile as organic modifier. (1) benzylal-cohol; (2) acetophenone; (3) phenylethanol; (4) propiophenone; (5) anisole; (6) toluene; and (7) p-cresol*

of mass spectrometers as routine detectors in LC, there has been a change toward the use of volatile buffers of organic acids such as acetic acid or formic acid and their ammonium salts. These buffer substances also have a better solubility in organic solvents.

When performing separation of acids or bases, a change in pH of the mobile phase can cause large changes in separation selectivity. Figure 4.19 shows how the retention of acids and bases varies with pH. Retention decreases with increasing ionization. If an acid in the sample has a pK_a value of 4.5, it is 50% ionized at pH 4.5 and fully ionized at pH above 6.5. The partition between the mobile and stationary phases is at high pH shifted toward the mobile phase, resulting in less retention. Retention varies greatly in the range around the pK_a value, where small changes in pH provide major changes in ionization and in retention. The weak acid is retarded most strongly when the ionization is suppressed at low pH. The same consideration is true for a weak base at high pH, where the ionization is suppressed. If it has a pK_a value of 9.5, half of the molecules will be ionized at pH 9.5, and at pH below 7.5 it will be fully ionized. Many silica-based column packing materials cannot be operated above pH 8, and therefore amines are often chromatographed in the pH interval of 3–7. Retention is then unaffected by pH as long as the base is fully ionized. In the pH range where either drugs are fully ionized or the ionization is completely suppressed, the retention is not affected by minor changes in pH.

To obtain robust analytical methods, it is important to choose a pH where retention varies only little with small changes in the composition of the mobile phase. It is unfavorable to

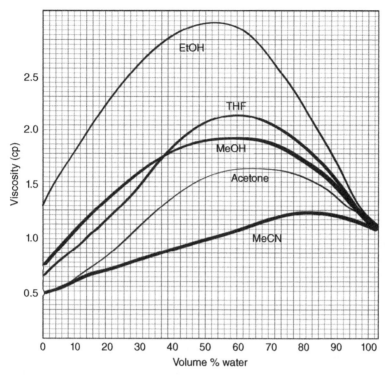

Figure 4.18 *Viscosity of mixtures of water and organic solvents. EtOH, ethanol; THF, tetrahydrofuran; MeOH, methanol; MeCN, acetonitrile*

Table 4.4 *pKa values (25°C) of buffer compounds frequently used in HPLC*

Buffer	pKa1	pKa2	pKa3
Acetate	4.75	—	—
Ammonium	9.24	—	—
Borate	9.24	12.74	13.80
Citrate	3.15	4.75	6.40
Ethylenediamine	7.56	10.71	—
Formate	3.75	—	—
Glycine	2.35	9.78	—
MES	6.15	—	—
Oxalate	1.27	4.27	—
Phosphate	2.14	7.20	12.35
Trichloroacetic acid	0.52	—	—
Triethanolamine	7.76	—	—
Triethylamine	10.72	—	—
Trifluoroacetic acid	0.50	—	—
TRIS	8.08	—	—

HPLC, high-performance liquid chromatography; MES: 2-(*N*-morpholino)ethanesulfonic acid; TRIS: tris(hydroxymethyl)aminomethane.

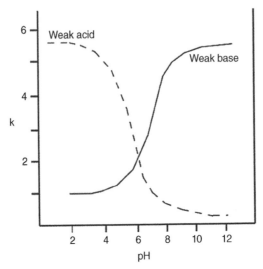

Figure 4.19   *Variation of the retention factor, k, for a weak acid and a weak base as a function of pH in the mobile phase*

select the pH near the pK_a value of the analytes, unless it is a question of the separation of two closely related substances with a small difference in pK_a values. Typically, acids are separated by a pH that suppresses ionization and bases at a pH where they are ionized. When using column packing materials stable in the entire pH range, it is of course possible to chromatograph bases in the ion suppressed mode. With the development of stationary phases to be used also at high pH, new possibilities in selectivity changes have become available (Figure 4.20).

4.10.4   Ion Pair Chromatography

In Section 4.10.3, it was recommended to separate the substances at a pH that suppresses the ionization or at a pH where the substances are completely ionized. In the latter case, the retention may be too short. In that case, *ion pair chromatography* should be considered. The technique is particularly useful for separation of basic compounds in reversed phase chromatography.

In ion pair chromatography, the analytes have to be ionized, and therefore pH has to be controlled. The mobile phase typically has a pH of 7 where both carboxylic acids and aliphatic amines are ionized. All ions in the system are surrounded by ions of opposite charge in order to keep electro neutrality in the system. Buffer ions are normally very polar and have only little interaction with the analytes. But when larger ions are added to the mobile phase, they can form ion pairs with other ions of opposite charge in the system. In ion pair chromatography, larger hydrophobic ions are added to the mobile phase. The formed hydrophobic ion pairs appear neutral to the surroundings. The ion-pairing process is a dynamic process where molecules are exchanging all the time but when adding the ion-pairing reagent, the *counter-ion*, in excess, there will be a high probability for the

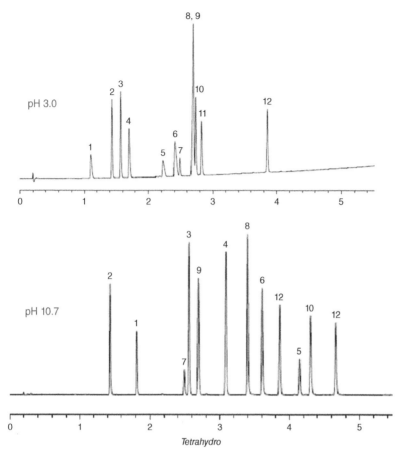

Figure 4.20 *Separation selectivity at high and low pH. Separation is achieved using gradient elution on an ACE UltraCore 2.5 μm Super C18, 50 mm × 2.1 mm column. (a) At pH 3.0, an acetonitrile gradient containing ammonium formate pH 3.0 was used. (b) At pH 10.7, an acetonitrile gradient containing 18 mM ammonia pH 10.7 was used. Sample: (1) atenolol; (2) methylphenylsulfoxide; (3) eserine; (4) prilocaine; (5) bupivacaine; (6) tetracaine; (7) 1,2,3,4-tetrahydro-1-naphthol; (8) carvedilol; (9) nitrobenzene; (10) methdilazine; (11) amitriptyline; and (12) valerophenone. Reproduced with permission of Advanced Chromatography Technologies Ltd, UK (info@ace-hplc.com/www.ace-hplc.com)*

formation of an ion pair with the analyte. The ion pair is hydrophobic and will have a retention depending on the nature and concentration of the counter-ion.

When the molecular size or the concentration of the counter-ion increases, the retention will increase. However, the effect of increasing concentration of the counter-ion decreases at higher concentrations and therefore concentrations in the range of 1–10 mM are recommended.

Sulfonic acids and perfluorocarboxylic acids are common counter-ions for protonated bases, and quaternary ammonium compounds are commonly used for ionized acids.

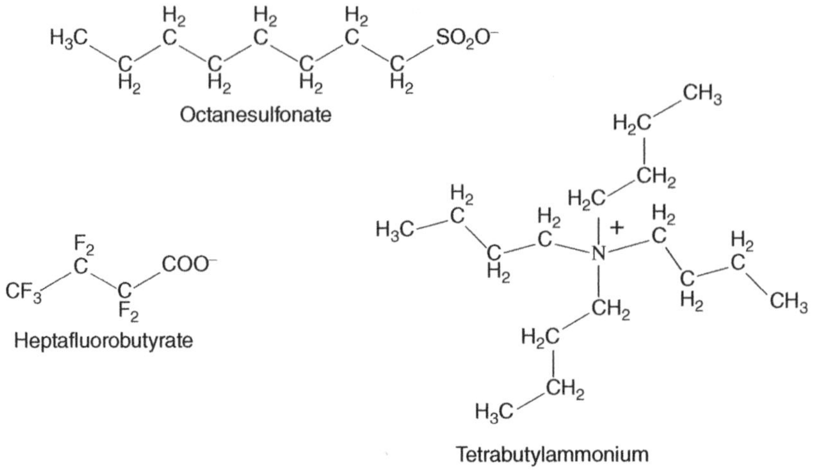

Figure 4.21   *Structures of the ions octanesulfonate, heptafluorobutyrate, and tetrabutylammonium*

Figure 4.21 shows the structure of three typical counter-ions: octanesulfonate, heptafluorobutyrate, and tetrabutylammonium. Octanesulfonic acid has a pK_a below 1, is negatively charged throughout the used pH range, and forms ion pairs with positively charged bases. Heptafluorobutyrate behaves similarly and can also be considered negatively charged throughout the pH range. Tetrabutylammonium is a quaternary ammonium compound that is positively charged in the full pH range. The positively charged group provides ion pairs with negatively charged acid groups.

Octanesulfonate provides larger retention of bases than pentanesulfonate, and tetrabutylammonium ions provide larger retention of acids compared to tetrametylammonium ions. Retention in ion pair chromatography is increased by changing the mobile phase as follows:

- Reduce the concentration of organic modifier.
- Increase the concentration of the counter-ion.
- Increase the molecular size of the counter-ion.

A drawback of ion pair chromatography is that many ion-pairing reagents are nonvolatile and thus not compatible with MS detection. Only the fluoro-carboxylic acids are sufficiently volatile to be used together with MS detection.

4.11   Size Exclusion Chromatography (SEC)

Size exclusion chromatography (SEC) is a separation method in which substances are separated according to their molecular size. The principle of SEC is based on a sieving effect and not on partition between immiscible phases. In this section, we shall look briefly into the separation principle.

4.11.1   Principle

In SEC, analytes are separated according to their molecular size in solution. This technique is primarily used for separation of larger molecules. However, it is also sometimes used in sample preparation in order to separate small molecules from larger molecules. It is very important that the analytes are very soluble in the mobile phase used. In this separation system, the stationary phase is identical with the mobile phase, and the stationary phase is held within the pores of a totally porous matrix. This means that the liquid phase surrounding the matrix particles is identical with the stationary phase inside the pores of the matrix particles. To obtain true SEC, the analytes must not interact with the matrix particles. If interaction between the analytes and the matrix particles takes place, it can be considered as a kind of adsorption chromatography. This will compromise the SEC analysis.

In order to separate hydrophobic substances, organic solvents are used as mobile phases and the technique is called *gel permeation chromatography* (GPC). For water-soluble substances, aqueous mobile phases are used and the technique is called *gel filtration chromatography*. Separation occurs when analytes penetrate the pores of the matrix. The more of the pore volume the analytes "see," the more they will be retained. Analytes that can penetrate to the full pore volume will be eluted corresponding to the total volume of liquid in the column. This volume corresponds to the holdup volume in a standard LC system. This is illustrated in Figure 11.14. Molecules not able to penetrate the pores will be eluted corresponding to the volume of liquid surrounding the matrix particles.

The large molecules pass through the column outside of the particles. The volume of mobile phase that has carried them through the column is called the *exclusion volume*, V_0. The exclusion volume is the volume of mobile phase between the particles in the column. The smallest drug molecules are transported by the mobile phase into the smallest pores of the particles. Because the path length through all the pores of the packing material is much longer, the small molecules get a longer retention time. The volume of mobile phase that is used to elute these substances is called *the total permeation volume*, V_M. Figure 4.22 shows a chromatogram and a calibration curve of molecular mass versus retention. The retention increases with decreasing molecular mass. All molecules in a sample will be eluted within a volume between V_0 and V_M as long as no interaction with the matrix particles takes place.

The matrix particles for SEC are manufactured with controlled pore sizes. The pore size is chosen dependent on the molecular range to be investigated. Because all analytes are eluted between the exclusion volume and the total holdup volume of the column, only a limited PC is obtainable. Therefore, it is important to choose a column with matrix particles that have pore sizes suitable for the molecules in the sample to be separated. A given SEC column has a given molecular weight range for fractionation. If the fractionation range for proteins is given as 30 000–200 000 Da, it means that proteins larger than 200 000 Da are excluded from the pores and thus all will elute with the exclusion volume. All substances smaller than 30 000 Da will fully penetrate all the pores, get maximum retention, and elute together in the holdup volume (the total permeation volume). The column can thus separate substances with molecular weight between 30 000 and 200 000 Da. Thus, large molecules that do not enter the pores cannot be separated from each other, and analytes that are so small that they all penetrates all pores cannot be separated.

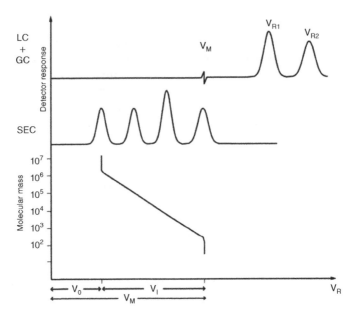

Figure 4.22   *Liquid chromatography (LC) versus size exclusion chromatography (SEC) and the calibration curve for SEC*

Packing materials can be soft or hard. Soft materials are made of polysaccharides such as dextran, polyacrylamide, or polystyrene. Soft gels are used with gravity flow (flow driven only by the gravity), where the mobile phase flows through the column as a function of gravity only. For HPLC, rigid packing materials that resist the higher pressure are used. Rigid packaging materials are made of silica or of a highly cross-linked organic polymer like PS-DVB co-polymer. Silica is used with aqueous mobile phases, whereas PS-DVB often is used with organic solvents as mobile phases.

When molecular weight is to be determined by SEC, it is important to realize that the size of large molecules like proteins or polysaccharides depends on the environment they are dissolved in. A coiled molecule may change it size due to its concentration, the ion strength of the solvent, solvation, and so on. Calibration standards with known molecular weight have to be of a similar structure as the analytes to be determined. Otherwise, there is a risk of a major bias on the result. The calibration curve is a logarithmic function, and small variations in elution volume will lead to major errors in the molecular weight determined. Thus, the accuracy of the molecular weight determination by SEC is not impressive, and the technique is therefore more suited for characterization of polymers than for actual mass determination. The selectivity in SEC can be varied by changing the pore size distribution of the column packing material.

4.12   Ion Exchange Chromatography

Ion exchange chromatography is a technique that allows the separation of *ions* based on their charge. It can be used for almost any kind of charged molecules, including large *proteins*, *nucleotides*, and *amino acids*. The ionic analytes are retained by ionic interaction between the analytes and ionic sites of the opposite charge placed on the stationary phase. Ionic groups like $-SO_3^{2-}$, $-COO^-$, $-NH_3^+$, or $-NR_3^+$ are placed on the stationary phase particles. The charges are neutralized by counter ions, which can exchange with analytes. The positive cations will have affinity to negatively charged ion exchangers; these ion exchangers are therefore called *cation exchangers*, and positively charged ion exchangers are likewise called *anion exchangers*.

The ion exchanger can be classified into strong and weak ion exchangers, where the *strong ion exchangers* have either sulfonic acid groups, $-SO_3^{2-}$, or quaternary ammonium groups, $-NR_3^+$, attached to the stationary phase, and the *weak ion exchangers* have carboxylic acid groups, $-COO^-$, or an aliphatic amino group, $-NH_3^+$, attached. The aliphatic amino group can be a primary, secondary, or tertiary amine. The strong ion exchangers are charged throughout the usable pH range, whereas the weak exchangers only are charged in a minor range of the pH interval depending on the pK_a value of the functional group. In daily work, a *strong cation exchanger* is abbreviated as SCX, and a *strong anion exchanger* is abbreviated as SAX. Similarly, weak ion exchangers are called WCX or WAX.

The ionic adsorption sites on the column packing material are very specific, and therefore the column packing materials have a given capacity. The capacity is between 1 and 4 meq/g. A capacity of 4 meq/g of cation exchanger means that 1 g maximum can bind 4 mM of a protonated tertiary amine. It is important to realize that it is not possible to use the full capacity of a column when samples are injected. Only amounts of analytes of 5–10% of the capacity should be loaded onto such a system per injection in order to avoid overloading of the stationary phase.

Retention of analytes on an ion exchanger requires that analytes as well as the ion exchanger are charged. In general, the affinity of analytes toward the stationary phase depends on their charge and size. More charges and smaller hydrated size will increase the affinity for the ion exchanger and thus the retention. The retention can be changed by modifying the mobile phase either by addition of an organic solvent or more effectively by changing the concentration or nature of the buffer. The buffer ions in the mobile phase compete with the analyte ions for the ionic sites on the stationary phase. When the buffer concentration is increased or buffer ions with a stronger affinity to the ionic sites are used, the retention of the analytes will decrease. In cation exchange, the eluting strength of the buffer cation is increasing in the order Li^+, H^+, Na^+, NH_4^+, K^+, Ag^+, Mg^{2+}, and Zn^{2+}. In anion exchange, the eluting strength of buffer anions is increasing in the order OH^-, CH_3COO^-, $HCOO^-$, Cl^-, Br^-, $H_2PO_4^-$, oxalate, and citrate. These are, of course, only general rules and may change with the type of ion exchanger. It is common to perform gradient elution in ion exchange chromatography by increasing the ionic strength of the buffer during chromatography.

4.13 Chiral Separations

In Chapter 2, a short discussion of stereochemistry was given. From this, it is evident that it is a challenge to separate pairs of enantiomers as they have the same physicochemical characteristics apart from their ability to rotate the plane-polarized light. If two substances have the same distribution constants, they cannot be separated in a chromatographic system.

Two enantiomers can therefore only be separated in chromatographic systems if it is possible to introduce a difference in their distribution constants. This can in principle be done in two ways: an indirect way and a direct way.

In the indirect way, the enantiomers are derivatized with an optically active reagent to form diastereomers. Diastereomers have more than one chiral center and will have different physicochemical characteristics and thus also different distribution constants. It is therefore possible to separate enantiomers after derivatization to diastereomers in a standard chromatographic system. This can be by either LC or GC.

When performing derivatization with chiral reagents, it is important to be aware of the purity of the derivatization reagent. The analyte (A), being a mixture of the R- and S-enantiomer of the analyte (e.g., 1% of one in the other), is derivatized with the optical active reagent (R). The ideal derivatization process with a 100% pure reagent (R-R) will result in two diastereomers:

$$\text{Ideal derivatization process}: \text{R, S-A} + \text{R-R} \rightarrow \text{R-A, R-R} + \text{S-A, R-R}$$

However, if (when) the reagent contains an enantiomeric impurity, the reaction looks like this:

$$\text{Derivatization process}:$$
$$\text{R, S-A} + \text{R, S-R} \rightarrow \text{R-A, R-R} + \text{S-A, S-R} + \text{S-A, R-R} + \text{R-A, S-R}$$

The four different substances achieved will in pairs be mirror images of each other (enantiomers) and thus result in only two chromatographic peaks. If an analyte contains 1% of the enantiomer and it is analyzed by chiral derivatization followed by achiral HPLC analysis using a chiral derivatization reagent containing 1% of its enantiomer, the bias of the analysis will be about 100%, as shown in Table 4.5.

If the derivatization results in a given order of elution of the two peaks, it will be possible to reverse this order of elution if the other chiral enantiomer of the reagent is used for derivatization. In this way, it is possible to elute the minor peak before the major peak, avoiding that the minor peak elutes as a shoulder on the tail of the major peak, and this will in many cases improve the quality of the quantitative measurement.

Table 4.5 *Derivatization of an enantiomeric analyte with a chiral reagent*

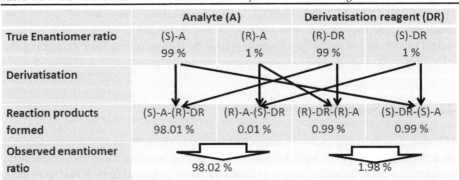

	Analyte (A)		Derivatisation reagent (DR)	
True Enantiomer ratio	(S)-A	(R)-A	(R)-DR	(S)-DR
	99 %	1 %	99 %	1 %
Derivatisation				
Reaction products formed	(S)-A-(R)-DR	(R)-A-(S)-DR	(R)-DR-(R)-A	(S)-DR-(S)-A
	98.01 %	0.01 %	0.99 %	0.99 %
Observed enantiomer ratio		98.02 %		1.98 %

In bioanalysis, the task is to investigate whether any chiral inversion takes place in vivo. It can be a drug given as a racemate and where some inversion may take place in vivo, or it can be a drug given as one of the enantiomers where any inversion is to be controlled.

When using direct separation of enantiomers, they are injected directly into the chromatographic system similar to what is used in other chromatographic methods. The possible bias from derivatization is thus avoided. The direct chiral separation (without derivatization) of enantiomers is only possible if chirality is introduced into the chromatographic system. Chirality of chromatographic systems is achieved using chiral mobile or stationary phases. In GC, it is only possible to have a chiral stationary phase, and this can be performed using either phases based on cyclodextrins or phases based on polymers into which optically active amino acids have been incorporated. In LC, chirality can be introduced in the mobile or in the stationary phase, and both modes have been applied. If chirality is in the mobile phase, diastereomeric complexes with different complex constants have to be formed to achieve separation on an achiral stationary phase. A chiral reagent is added to the mobile phase, and due to the different complex constants of the diastereomeric complexes formed with the enantiomers to be separated they can be separated on an achiral stationary phase.

But, more often, chiral stationary phases have been used, and a huge number is available on the marked. The phases consist of a polymer (silica, cellulose, or methacrylate) to which a chiral molecule is attached. The chiral molecules used are very different in nature (proteins, polysaccharides, cyclodextrins, antibiotics, helical methacrylates, etc.) and will therefore also provide different selectivity. Unfortunately, achievement of chiral separation is still much of a trial-and-error process, and therefore it is important to consult literature and vendors of stationary phases before starting the experiments.

5

Quantitative and Qualitative Chromatographic Analysis

Steen Honoré Hansen

School of Pharmaceutical Sciences, University of Copenhagen, Denmark

This chapter describes how to perform quantitative measurement using chromatographic separation techniques. Different methods of calibration such as the external standard method, the internal standard (IS) method, the standard addition method, and normalization are discussed. Also, a short introduction to qualitative analysis using chromatographic separation techniques combined with mass spectrometry and nuclear magnetic resonance spectroscopy is given.

5.1 Collection of Chromatographic Data

The data used for the quantitative calculations are obtained from the chromatogram, being a result of the chromatographic separation. It is therefore important that the chromatogram is of high quality in order to eliminate uncertainty. Ideally, the chromatogram is without any baseline noise and only show peaks fully separated to baseline. However, this is seldom the case in bioanalysis, where samples consist of thousands of compounds.

The chromatographic baseline always shows a certain noise level coming from the instrument (from the electronics as well as pump noise; Figure 5.1). In order to obtain a reliable measurement of the chromatographic peak, a *signal-to-noise level* should as a rule of thumb be at least 10 (Figure 5.1).

The peak height (H) must be 10 times the peak-to-peak noise (h) at the baseline. This is called the *limit of quantification* (LOQ) or the *lower limit of quantification* (LLOQ). Sometimes, the *limit of detection* (LOD) is also mentioned, and this is typically defined as H being two or three times the peak-to-peak noise. Both the LOD and the LOQ are dependent on the current status of the instrument, and if the peak-to-peak noise level changes so will

Bioanalysis of Pharmaceuticals: Sample Preparation, Separation Techniques and Mass Spectrometry,
First Edition. Steen Honoré Hansen and Stig Pedersen-Bjergaard.
© 2015 John Wiley & Sons, Ltd. Published 2015 by John Wiley & Sons, Ltd.

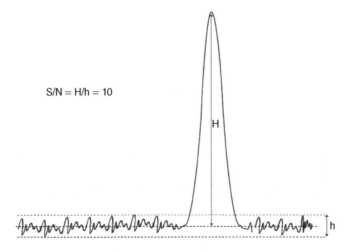

S/N = H/h = 10

H

h

Figure 5.1   *A chromatographic peak at the limit of quantification*

the LOD and the LOQ. From Figure 5.1, it is obvious that the peak height or peak area will be measured with a certain uncertainty depending on how the baseline is defined. It is therefore necessary to validate the method in order to show that the *coefficient of variation* (CV) or the *relative standard deviation* is within 20% at the LOQ concentration level. Validation is performed according to international guidelines (see Chapter 11).

5.2   Quantitative Measurements

Quantitative measurements using chromatographic analysis are based on measurements of peak height or peak area of the peak from a sample with the unknown concentration, and comparing this with a corresponding measurement of a calibration standard of known concentration. Thus, chromatographic analysis is not an absolute measurement but is traceable to a traceable and well-defined calibration standard.

The measurement of peak heights can be made with high precision and yields accurate results, provided variations in column conditions do not alter peak width. As peak height is inversely related to peak width, all parameters that can affect the peak width must be held constant. The parameters that affect peak height are the retention factor, the number of theoretical plates, and the symmetry factor. When the retention factor increases, peak broadening occurs and the peak height decreases. A reduction in the number of theoretical plates and an increase of the symmetry factor also decrease peak heights.

Peak area is virtually independent of peak-broadening effects. From this standpoint, therefore, peak area is a more satisfactory analytical parameter than peak height. In modern instruments, peak area is measured with a high degree of *accuracy* and *precision* also if the peak is not symmetrical, and the peak area is usually the recommended parameter for quantitative determinations.

Peak heights and peak areas are dependent on the properties of the analyte, on the properties of the detector, and to some extent also on the constituents of the matrix. To determine unknown concentrations, calibration curves must be prepared for each analyte and preferably in the same matrix as the sample. The most important calibration methods are:

- External standard method
- Internal standard method
- Standard addition
- Normalization.

5.3   Calibration Methods

5.3.1   External Standard Method

An external standard is a certified reference substance of the compound to be determined or a substance that has similar high quality. This means that it has to be pure, or impurities must not have the same retention as the analyte or any internal standard. The concentration has to be known, and the substance has to be stable.

In the external standard method, a series of standard solutions containing known concentrations of the reference substance are prepared separately in a blank (drug-free) matrix (plasma, urine, etc.). These solutions are called external standards solutions, and their concentrations approximate the concentration range of the unknowns. The standard solutions are analyzed separately from unknown samples under identical conditions. The peak areas or the peak heights of the reference standards are read from the chromatograms and are plotted as a function of concentration. The *calibration curve* should preferably yield a straight line, but concave or even convex calibration curves are sometimes experienced. Quantitative determination at the LOQ level is often necessary in bioanalysis when the data are to be used for pharmacokinetic calculations. One problem often encountered with calibrations curves is that even if they have been tested with linear regression and a coefficient of determination (r^2) is found close to 1, the calibration points at the lower part of the calibration curve may show deviation from the linear curve—often showing too high results due to interferences or carryover in the system. To verify such deviations, a residual plot of data should be performed. If the deviation is confirmed, some data weighting should be considered.

It is a good idea to include *quality control samples* (QC samples) in each batch run of samples when analyzing larger numbers of the same samples. The QC samples are prepared in a sufficient volume at two or three concentration levels by spiking blank matrix. Each QC sample is then divided into a number of vials and kept frozen until use. A number of QC samples are then included in each batch run of unknown samples to keep track of the quality of the analysis.

The calibration solution should cover the concentration range from LLOQ to the *upper limit of quantification* (ULOQ) (Figure 5.2). Data obtained being outside (above) this calibration range are not reliable, and sample should be diluted and reanalyzed. Normally, calibration standard solutions are prepared in the same matrix as the sample exists in. But some biofluids and tissue are fairly different from sample to sample (e.g., urine), and this may increase the uncertainty of the data obtained. To compensate for this, internal standardization can be used.

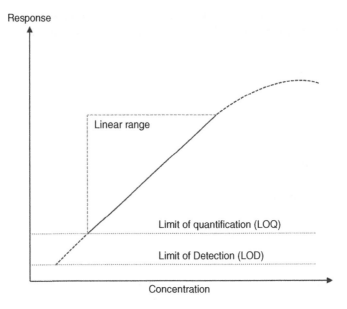

Figure 5.2   *An example of a quantitative determination calibration curve*

5.3.2   Internal Standard Method

The use of the internal standard method may compensate for analytical errors due to sample losses during sample preparation and variable injection volumes. The method is similar to the external standard method in that solutions of the reference standard are compared with solutions of the sample. The key difference is that a compound called an *internal standard* is added at an identical concentration to all standard solutions and to all sample solutions prior to any sample pretreatment. After sample processing and injection into the chromatographic system, the peaks of the analyte and the internal standard are shown in the chromatogram, as illustrated in Figure 5.3.

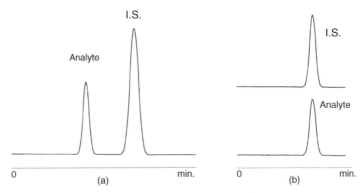

Figure 5.3   *Chromatograms showing the use of internal standards (IS). (a) An example where the internal standard is separated from the analyte. (b) An example where the internal standard is an isotopically labeled analyte and the two compounds are measured by mass spectrometry (MS) at to different mass-to-charge ratios (m/z) values*

The analyte or internal standard peak area or peak height ratio is used in the calibration rather than the absolute values such as in the external standard method. The internal standard method can improve precision when the dominant sources of error are related to sample preparation or injection. Such errors affect both the internal standard and the analyte peak in the same way, and the relative peak area or peak height ratios become unaffected. For this method to work well, it is important to choose a suitable internal standard. The internal standard should match the analyte closely. Ideally, an internal standard should:

- be recovered to a similar extent as the analyte in the sample preparation procedure (i.e., have the same physicochemical behavior as the analyte (log P, detectability, vapor pressure, etc.).
- be separated from all other substances in the sample.
- have a retention time and a detector response that is similar to those of the analyte.
- have a concentration that gives a similar peak height or area as that of the analyte.
- be stable.
- On one hand, the internal standard does not have to be pure; but, on the other, it must not contain any impurity that may interfere with the analyte, and the concentration does not have to be known.

Consequently, the internal standard should have very similar physicochemical properties as the analyte, but not so similar that it cannot be accurately determined. The ideal internal standard in liquid chromatography–mass spectrometry (LC-MS) or gas chromatography–mass spectrometry (GC-MS) is an isotopically labeled version of the analyte. A labeled internal standard matches the analyte and is accurately determined at a higher mass-to-charge ratio (m/z). Typically, 5–8 deuterium atoms are introduced for hydrogen, but also ^{13}C and other isotopes can be used. It is important that the m/z value for the internal standard does not interfere with the isotope peaks of the analyte. Labeled internal standards are expensive, and therefore in many situations compound analogs are used as internal standards. Many drug substances are tertiary amines. An analog compound containing an extra CH_2 group could be a possible IS, whereas a corresponding secondary amine would be a bad choice because tertiary and secondary amines behave differently. The internal standard method is a key method in bioanalysis.

In the internal standard method, a series of standard solutions containing known concentrations of the reference standard are prepared in the actual matrix. Standard solutions for determinations of drug substances in plasma are prepared in drug-free plasma, and standard solutions for determination of drug substances in urine are prepared in drug-free urine. Then, the same amount of internal standard is added to all standard solutions and to all sample solutions. All solutions are treated in an identical way in the sample preparation procedure and are subsequently injected into the column. The peak areas of the analyte and the internal standard are read from the chromatograms, and the analyte/internal standard peak area ratios in each solution are calculated. The peak area ratios of standard solutions are plotted as a function of concentration. The calibration curve should preferably yield a straight line. QC samples are also used in methods with internal standardization.

5.3.3   Standard Addition

Standard addition is a calibration method used to determine the concentration of an analyte in a complex analytical matrix that contains substances that may interfere with the detector

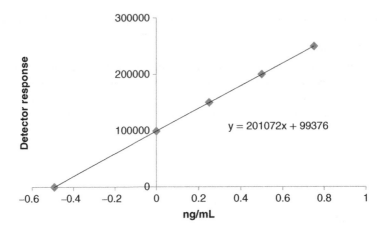

Figure 5.4   *The standard addition calibration curve*

response of the analyte. Because of interference, a calibration curve based on pure analyte samples will give an incorrect determination. If such interference varies from sample to sample, compensation can be achieved using isotopically labeled analyte because such a compound will be affected in the same way as the analyte. But if no isotopically labeled analyte is available, the method of standard addition can be applied.

Adding known concentrations of a chemical reference standard of the analyte to the unknown solution can solve the problem. This calibration principle is called standard addition. An example of an obtained "calibration curve" for a given sample using standard addition is shown in Figure 5.4. The curve is constructed from data obtained the following way.

The concentration of an unknown sample solution is expected to be about 1.0 ng/ml. To determine the unknown concentration, four standard addition solutions (SASs) were prepared from the sample solution. Volumes of 5 ml were taken from the sample solution and added to four 10 ml volumetric flasks. SAS #1 was prepared by diluting 5 ml of the sample solution to 10 ml. SAS #2, 3, and 4 were prepared by mixing 5 ml of the sample solution with, respectively, 0.25, 0.5, and 0.75 ml of a reference solution followed by dilution to 10 ml. The reference solution contained 10 ng/ml of a chemical reference standard of the analyte. The detector responses, which are the sum of the unknown concentration (x) and added concentrations, are then plotted versus the detector response.

The regression line is calculated according to the method of least squares using the four data points, and the content can be calculated for $y = 0$. This will result in $x = 0.494$, and, taking dilution into consideration, the sample contains 0.99 ng/ml.

5.3.4   Normalization

Normalization is a technique used to provide a quantitative analysis of a mixture that is separated by a chromatographic method. It can be applied to peak areas (but not to peak heights), and it is assumed that the detector response (detector signal per gram of analyte) is the same for all components of the mixture. The quantitative results are obtained by expressing the area of a given peak (A) as a percentage of the sum of all peak areas of the components (A$_i$) in the mixture:

$$A(\%) = \frac{A}{\sum Ai} \times 100 \qquad (5.1)$$

5.4   Validation

The analytical method should always be validated before use (Figure 5.5). A number of international guidelines are available, and they are fairly similar in requirements. One example of such a guideline is the Food and Drug Administration (FDA) "Guidance for Industry: Bioanalytical Method Validation." This is further discussed in Chapter 11.

5.4.1   Accuracy and Precision

The accuracy of an analytical procedure is defined as the closeness of the test results obtained by the procedure to the true value. The accuracy is estimated by measuring spiked samples according to the analytical procedure and comparing these results with results obtained by analyzing calibration standards directly. The accuracy determined toward calibration standards made up in water (the recovery) should preferably be high in order to have a robust method. However, when performing analysis of biofluids, it is

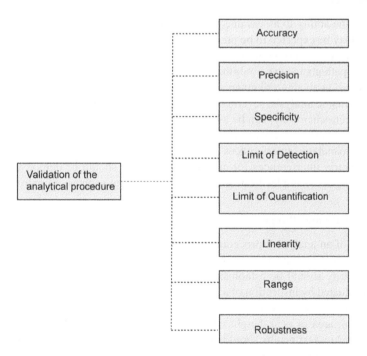

Figure 5.5   *Parameters to be validated in a bioanalytical method*

common to use calibration standards spiked in the same matrix as the samples. In this way, a recovery less than 100% is compensated when calibration standards and samples are processed in an identical way.

The accuracy of the analytical procedure should be established across its entire calibration range. It is recommended to use three concentration levels: one a little above LLOQ, one in the middle of the range, and one close to the ULOQ. At each of these three levels, six replicates applying the full analytical procedure are performed. The relative standard deviation (% RSD) at each level expresses the precision of the method.

When the precision is determined within a short interval of time (e.g., on the same day), it is called *repeatability* or *precision within day*. The measurements are often repeated on several subsequent days, and the overall precision is called *precision between days*. When the *reproducibility* of the method has to be determined, it is necessary to involve more than one laboratory.

5.4.2   LOD and LLOQ

LOD is only a value that tells the reader what the limit is with respect to detecting the analyte with a certain probability. LOD is not to be used in quantitative analysis because the data are too unreliable. The requirement for LLOQ is that the precision should be within 20% RSD. At all other concentrations, the precision should be within 15% RSD.

5.4.3   Specificity

Specificity is the ability to assess unequivocally the analyte in the presence of other compounds that may be expected to be present. Sometimes, a number of other drug substances are screened to verify any interference. However, it is not possible to test all substances; and, as some patients receive a number of drugs, the only way to obtain a high degree of reliability when analyzing unknown samples is to use more than one analytical method. In chromatographic analysis, several selective detectors (e.g., mass spectrometry, fluorescence, and electrochemistry) can be used.

Selectivity studies should also assess interferences that may be caused by matrix components. In bioanalysis, interferences caused by biological fluids must be examined. The absence of matrix interferences for quantitative methods should be demonstrated by the analysis of drug-free samples from at least five independent sources of control matrix.

5.4.4   Linearity and Range

The *linearity* of an analytical procedure is its ability (within a given range) to obtain test results, which are directly proportional to the concentration (amount) of analyte in the sample. Linearity should be established across the *range* of the analytical procedure. It should be evaluated by inspection of a plot of signals as a function of analyte concentration or content. The signal can be absorbance in a spectroscopic procedure and peak area, peak height, or peak area or peak height ratios in a chromatographic procedure. If there is a linear relationship, test results should be evaluated by appropriate statistical methods, for example by calculation of a regression line by the method of least squares. The square of the correlation coefficient and the regression line should be reported. It is important to realize that the calibration curve does not have to be strictly linear, but it has to be described by a suitable equation for the regression line to be able to use it for reliable quantitative measurements.

For establishing linearity, a minimum of five concentrations should normally be used. If the calibration curve is nonlinear, 8–10 concentrations should be used depending of the range. For bioanalysis, the range is normally established from the quantification limit and to somewhat above the highest expected concentration found in the samples. The range of an analytical procedure is the interval between the ULOQ and LLOQ of the analyte in the sample (including these concentrations) for which it has been demonstrated that the analytical procedure has suitable precision, accuracy, and linearity.

5.4.5 Robustness

The *robustness* of an analytical procedure is a measure of its capacity to remain unaffected by small but deliberate variations in method parameters, and it provides an indication of its reliability during normal usage. The evaluation of robustness should be considered during the development phase and depends on the type of procedure under study. In the case of LC, examples of typical variations are:

- influence of variations of pH in the mobile phase
- influence of variations in mobile phase composition
- different columns (different lots or suppliers)
- temperature
- flow rate.

In the case of GC, examples of typical variations are:

- different columns (different lots or suppliers)
- temperature
- flow rate.

5.5 Qualitative Analysis

The use of qualitative analysis in bioanalytical chemistry is typically concerned with the identification of organic molecules in samples of biological origin. This can be in relation to biomarker identification or elucidation of the structure of a metabolite.

The structure of a biomarker molecule can vary from a very small molecule to large biomolecules like proteins or nucleic acids. The process used to identify possible biomarkers involves a number of laborious tasks. Initially, pictures of all of the compounds in samples achieved from persons carrying a given disease are compared with corresponding samples achieved from healthy persons using omics techniques. This can be LC-MS for small molecules or 2D electrophoresis for proteins. When it comes to small molecules, differences are evaluated, and possible biomarker candidates are then identified after further fractionation using high-resolution mass spectrometry (HRMS) and nuclear magnetic resonance (NMR). For proteins, LC fractionation or 2D electrophoresis is combined with MALDI-TOF-MS (matrix-assisted laser desorption/ionization time-of-flight mass spectrometry) and other proteomics techniques (e.g., sequencing of peptides obtained by tryptic digest of the protein).

In drug discovery, it is important to know the metabolites formed in the different animal species and in humans used in the R&D process. Any metabolite formed in humans should also be present in the animal species used for the toxicology studies in order to secure that no toxic side effect can be related to new human metabolites. Metabolite identification of a small-molecule drug is a little more straightforward compared to biomarker identification. The advantage is that the structure of the parent drug is known in advance. The question is therefore only how to detect any transformation products coming from this molecule. If the molecule contains a special structure element that is not found in endogenous molecules (e.g., a bromine atom), and that is not prone to any metabolic or other changes and can be selectively measured, it can be used to trace any metabolites formed. If such a handle is not present, it must be introduced into the molecule. Isotope labeling of the drug molecule is a common way to solve this problem. The labeling is most often performed with radio isotopes like ^{3}H or ^{14}C because radioactivity detection is sensitive and specific. ^{14}C labeling is preferred because tritium more or less readily may exchange with protons. It is therefore important that the labeling is incorporated into the molecule at a position where no transformation will take place. If the drug is cleaved into two parts, it may be necessary to include more than one label.

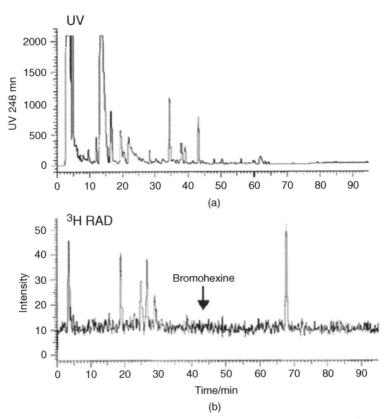

Figure 5.6 *Chromatograms of bromohexine and its metabolites in urine using (a) ultraviolet and (b) radiochemical detection*

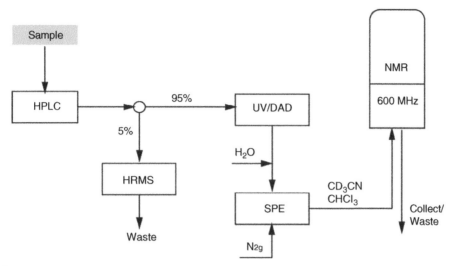

Figure 5.7 *High-performance liquid chromatography–ultraviolet–mass spectrometry–nuclear magnetic resonance (HPLC-UV-MS-NMR) system for metabolite identification*

Initial investigations of drug metabolism are performed in in vitro tests using microsomes or hepatocytes isolated from the respective animal species. Human microsomes are commercially available. The advantage of the in vitro test is that these samples contain fewer endogenous compounds compared to samples achieved from in vivo studies. Analysis is performed using LC-UV-MS and combined with radiochemical detection (Figure 5.6).

After one or more metabolites are found, they have to be identified with full structure elucidation. This means that data, including HRMS data as well as NMR data, have to be obtained. In principle, HRMS and NMR data can be obtained on-line in a simple setup (Figure 5.7). However, acquiring NMR data on-line in a flow system requires relatively high concentrations of the metabolite, and therefore some kind of isolation of the metabolite is useful. This will also make the use of 2D NMR and other NMR experiments possible.

Collection of the chromatographic peak representing the metabolite using on-line solid phase extraction (SPE) can be repeated a number of times from repeated chromatographic runs of the sample and by using the same SPE cartridge. Subsequently, the SPE cartridge is dried using nitrogen, and the analyte is eluted with a deuterated solvent to the NMR.

Another possibility is the use of preparative high-performance liquid chromatography (HPLC) followed by NMR. NMR microtechnique, involving an NMR tube with an effective volume of only 30 µl, requires only a few micrograms of analytes to give a ^{1}H-NMR spectrum and a few hundred micrograms to give a 1D ^{13}C-spectrum. This amount of substance may readily be obtained using an analytical column for the preparation. Using this off-line technique requires that the isolation process is evaluated after each step, for example by LC analysis to ensure that no degradation or oxidation of the analyte has taken place. Remember that the collected eluate from the LC is diluted, and concentration by evaporation may be required to be able to detect the isolated peak. After having recorded the NMR spectra, the sample should again be analyzed by LC to verify that the original identity is maintained.

6

Sample Preparation

Stig Pedersen-Bjergaard[1], Astrid Gjelstad[2], and Trine Grønhaug Halvorsen[2]

[1]*School of Pharmacy, University of Oslo, Norway*
School of Pharmaceutical Sciences, University of Copenhagen, Denmark
[2]*School of Pharmacy, University of Oslo, Norway*

Biological fluids like human plasma and whole blood are very complex samples and normally cannot be injected directly into an analytical instrument for analysis of a drug substance. For this reason, some type of sample preparation is required prior to analytical measurement. This chapter discusses the principles and fundamentals of some of the most important bioanalytical sample preparation techniques, namely, protein precipitation (PPT), liquid–liquid extraction (LLE), and solid-phase extraction (SPE). This chapter also discusses those cases where sample preparation can be circumvented. Finally, the chapter finishes with a short discussion of alternative sample preparation approaches, and a discussion about some of the new technologies emerging in the field of bioanalytical chemistry.

6.1 Why Is Sample Preparation Required?

As discussed in Chapter 3, most biological fluids are very complex samples, and determining low-molecular drug substances or high-molecular biopharmaceuticals in such samples by chromatography and mass spectrometry is challenging for several reasons:

1. Complex biofluids can contain one or more matrix substances that can suppress the signal for the target analyte during the analytical measurement.
2. Complex biofluids can contain one or more matrix substances that give a false positive response (interference) for the target analyte during the analytical measurement.
3. Complex biofluids contain matrix substances that can contaminate or deteriorate the analytical instrument.
4. The biofluid itself can be incompatible with the analytical instrument because it is aqueous.

Bioanalysis of Pharmaceuticals: Sample Preparation, Separation Techniques and Mass Spectrometry,
First Edition. Steen Honoré Hansen and Stig Pedersen-Bjergaard.
© 2015 John Wiley & Sons, Ltd. Published 2015 by John Wiley & Sons, Ltd.

5. The concentration of the target analyte can be too low for detection in the analytical instrument.

The bioanalytical scientist should never underestimate the importance of these challenges. Although the target analyte is physically separated from most other sample constituents during the chromatographic separation, and subsequently measured by mass spectrometry at a mass value specific for the target analyte, the challenges above may sacrifice the quality of the bioanalytical results. To avoid this, the bioanalytical scientist should always pay close attention to *sample preparation*, which is defined as the chemistry done with the sample prior to injection in the analytical instrument.

Without sample preparation, some matrix components may elute at the same retention time as the target analyte (challenge #1). This is especially challenging when working with liquid chromatography–mass spectrometry (LC-MS). Although the matrix components have molecular masses that are different from the target analyte and will not be detected by the mass spectrometer, their co-elution with the target analyte can affect the ionization of the latter. Thus, the measured signal is partially suppressed and does not give an accurate quantitative determination. This is termed *ion suppression*. Phospholipids in plasma and serum are one class of matrix components known to cause serious ion suppression. Ion suppression will be discussed in more detail in Chapter 7. With proper sample preparation, major matrix components can be removed, which is often termed *sample clean-up*, and ion suppression can be avoided.

In some cases, the sample may even contain matrix components with similar mass and retention time as the target analyte (challenge #2). Under such very unfavorable conditions, the matrix components can directly interfere with the detection and measurement of the target analyte. This may cause false identification of the analyte in the sample, or inaccurate quantitative results. Also, this type of challenge can be circumvented by proper sample preparation, where the interfering matrix components are removed.

Among the different biological fluids used for bioanalysis, blood samples are especially complex. Blood samples contain a broad range of proteins, and the proteins can damage and clog an LC or gas chromatography (GC) column (challenge #3) when blood samples are repeatedly injected. To circumvent this, the proteins have to be removed during sample preparation. With efficient removal of the proteins, a large number of samples can be injected and analyzed in the analytical instrument (LC) without any problems.

Biological fluids are all aqueous samples. Aqueous samples are basically compatible with LC, but, as mentioned above, matrix components can deteriorate the analytical instrument and should be removed to improve the compatibility. Thus, the clean-up of the sample during sample preparation also enhances the compatibility with the analytical instrument. With GC, the sample compatibility becomes even more important because injection of aqueous samples should be avoided (challenge #4). In such cases, the target analytes should be transferred to an organic solvent, which is compatible with GC. Transfer to an organic solvent is an important part of the sample preparation prior to GC.

Most low-molecular drugs and biopharmaceuticals are present at low concentrations in biological fluids. Typically, concentration levels are from 0.1 ng/ml to 1 µg/ml. Before the introduction of highly sensitive instrumentation for LC-MS and GC-MS (gas chromatography–mass spectrometry), it was difficult to measure the lowest drug concentrations (challenge #5). At that time, *pre-concentration* of target analytes was important to be able to measure the lowest drug concentrations. Thus, pre-concentration was an important part of traditional bioanalytical sample preparation 10 years or more back in time. Today, LC-MS and GC-MS instruments provide very high sensitivity, and pre-concentration has become less important for many bioanalytical applications. However, if a concentration step is included in the sample preparation, it is still an advantage as this may increase the reliability in detection.

In general, sampling and sample preparation constitute over 80% of the total analysis time, and these steps are highly important for the overall success of the bioanalytical measurement. In the rest of this chapter, different techniques and strategies for sample preparation will be discussed in detail.

6.2 What Are the Main Strategies?

The most important sample preparation techniques in the modern bioanalytical laboratory include the following:

- PPT
- LLE
- SPE
- Dilute and shoot (DAS).

These techniques will be discussed in detail in Sections 6.3, 6.4, 6.5, and 6.6. Section 6.7 will discuss sample preparation techniques that are currently in use; they may be alternatives to the four main strategies, and they are commercially available.

6.3 Protein Precipitation

6.3.1 Fundamentals of Protein Precipitation

The most commonly used biological fluid for bioanalysis is plasma (or serum). Plasma contains about 8% (w/w) of proteins, as discussed in Chapter 3. Direct injection of plasma into LC (or other analytical instruments) is not recommended because the proteins easily precipitate due to their contact with the organic solvents and buffer salts commonly used in the mobile phase. Thus, with precipitated proteins in the LC column, the performance will rapidly deteriorate unless the major content of proteins in the samples is removed. PPT is intended to remove proteins from plasma or serum. The principle of PPT is illustrated in Figure 6.1.

The plasma or serum sample is pipetted into a small centrifuge tube. The volume of plasma is typically in the range from 10 to 1 000 μl. Subsequently, acetonitrile is pipetted into the plasma, and plasma and acetonitrile are mixed for a short period of time. Acetonitrile is a polar organic solvent, and it is miscible with the aqueous plasma sample. Thus, plasma and acetonitrile form a one-phase mixture. When the proteins come in contact with acetonitrile, denaturization occurs and the proteins precipitate. Thus, acetonitrile serves as a *precipitant*. The mixture is then centrifuged, and the clear liquid above the precipitate, termed the *supernatant*, is collected for the final analysis by LC or by another analytical technique.

The centrifugation step is important. The use of high g-forces improves the separation of the precipitate and the supernatant, and therefore centrifugation is performed at high

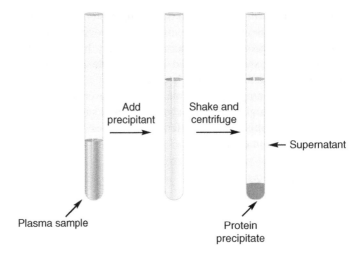

Figure 6.1   *Principle of protein precipitation*

rotations per minute, often equivalent to 10 000–15 000 g. To enable this, PPT and the following centrifugation are preferably performed in small centrifuge tubes especially made for strong centrifugation. Alternatively, the precipitated proteins can be removed by filtration, providing a clear *filtrate*, which can be injected into the analytical instrument. This will be discussed at the end of this section, when high-throughput PPT with 96-well technology is briefly discussed. It is important to emphasize that PPT is not an extraction technique. As discussed later in this chapter, extraction involves transfer of the target analyte from the sample medium into another medium. This is not the case in PPT, where the target analytes remain in the aqueous medium of the plasma sample.

6.3.2   The Most Frequently Used Precipitants

In the discussion related to Figure 6.1, acetonitrile was used as the precipitant. Acetonitrile is the most frequently used precipitant, but PPT can also be accomplished with several other precipitants, as illustrated in Table 6.1. In fact, there are four main approaches to PPT, namely, the addition of *organic solvent, acid, metal ions*, or a *salt* to the plasma sample. Among the different organic solvents used, acetonitrile, acetone, ethanol, and methanol are common. All are polar organic solvents, which are 100% miscible with aqueous plasma samples. As seen from Table 6.1, acetonitrile is the most efficient solvent, and the efficiency decreases in the order of acetonitrile > acetone > ethanol > methanol. Another very important observation from Table 6.1 is that the precipitation efficiency is dependent on the volume of precipitant added per volume of plasma. With acetonitrile as an example, 0.2 ml of acetonitrile added to 1.0 ml of plasma results in 13.4% protein removal. However, as the volume of acetonitrile is increased to 1.0 ml, the protein removal increases to about 97.2%. If the volume of acetonitrile is increased further to 2.0 ml, the protein removal is 99.7%, which gives a supernatant (or filtrate) essentially free from plasma proteins.

The organic solvents used as precipitants denature the plasma proteins. They are believed to act by decreasing the dielectric constant of the plasma proteins, which in turn increases the electrostatic interactions between the protein molecules. Also, the organic solvent

Table 6.1   *Different approaches to protein precipitation*

Precipitant	% Protein removal								
	Volume of precipitant added per volume of plasma								
	0.2	0.4	0.6	0.8	1.0	1.5	2.0	3.0	4.0
Organic solvent									
Acetonitrile	13.4	14.8	45.8	88.1	97.2	99.4	99.7	99.8	99.8
Acetone	1.5	7.4	33.6	71.0	96.2	99.1	99.4	99.2	99.1
Ethanol	10.1	11.2	41.7	74.8	91.4	96.3	98.3	99.1	99.3
Methanol	17.6	17.4	32.2	49.3	73.4	97.9	98.7	98.9	99.2
Acid									
Trichloroacetic acid (10% w/v)	99.7	99.3	99.6	99.5	99.5	99.7	99.8	99.8	99.8
Perchloric acid (6% w/v)	35.4	98.3	98.9	99.1	99.1	99.2	99.1	99.1	99.0
Metal ion									
$CuSO_4 + Na_2WO_4$	36.5	56.1	78.1	87.1	97.5	99.8	99.9	100.0	100.0
$ZnSO_4 + NaOH$	41.1	91.5	93.0	92.7	94.2	97.1	99.3	98.8	99.6
Salt									
$(NH_4)_2SO_4$ (saturated)	21.3	24.0	41.0	47.4	53.4	73.2	98.3	—	—

Data from Journal of Chromatography, 226 (1981) 455–460.

displaces water molecules associated with the hydrophobic regions of the proteins. Thus, both decrease of hydrophobic interactions and increase of electrostatic interactions result in protein aggregation and precipitation. Using organic solvents as precipitant is a relatively soft procedure, and this reduces the possibility for decomposition of labile analytes. On the other hand, if the supernatant is analyzed by LC or LC-MS, a high content of organic solvent may cause poor separation. This will be discussed in more detail in Chapter 7. A typical example of using acetonitrile for PPT is presented and discussed in Box 6.1.

Box 6.1 Protein Precipitation of Plasma Samples Prior to the Determination of Paroxetine

Background

Paroxetine is an antidepressant drug, and the typical concentration range of this substance in serum is 10–50 ng/ml. The structure of paroxetine is illustrated below.

Paroxetine is a basic substance with low water solubility. The pK_a-value is 9.8, and the log P-value is 3.15. In the procedure presented here, PPT is performed in plasma samples from patients treated with paroxetine, and the final analytical measurement (not discussed here) is by LC-MS.

Procedure

First, 250 µl of plasma is pipetted into a small centrifuge tube. Then, 25 µl of internal standard solution is added to the plasma. The internal standard is paroxetine-d_6, and the final concentration in the plasma sample of the internal standard is 100 ng/ml. Then, 500 µl of acetonitrile is added, and the mixture is vortexed and subsequently centrifuged at 10 000 g for 11 minutes. The supernatant is then transferred to a small glass vial for LC-MS analysis.

Fundamental Questions to the Procedure—Step by Step

1. Why is an internal standard added to the sample?
2. How much removal of proteins do you expect?
3. Why is the sample vortexed?
4. Why is the sample subjected to centrifugation?
5. Is the sample diluted?

Discussion of the Questions

1. The addition of an internal standard is performed to improve the precision and accuracy of the method. Thus, for the quantitative measurement, paroxetine is measured relative to the deuterated internal standard. This was discussed in Chapter 5.
2. The volume of plasma sample is 250 µl, and 500 µl acetonitrile is used as the precipitant. With a 2.0 volume of precipitant added per volume of plasma, the protein removal efficiency is expected to be 99.7% according to Table 6.1.
3. Vortexing (strong shaking) is performed to make sure that acetonitrile and plasma are mixed efficiently, to give as efficient precipitation as possible.
4. Strong centrifugation is performed to give a clear supernatant without any traces of precipitated protein and particulate matter.
5. The supernatant contains acetonitrile and water in the ratio of 2:1, and this is directly injected into LC-MS. The volume of supernatant is approximately 600 µl. As the original plasma volume was 250 µl, the PPT procedure results in dilution of the sample.

Alternatively, PPT can also be accomplished by addition of trichloroacetic acid (TCA) or perchloric acid to the plasma sample. Both reagents are highly efficient as shown in Table 6.1. One major advantage of using acids is that the supernatant (or filtrate) is still highly aqueous, unlike with acetonitrile where the supernatant typically becomes 50–66% organic. A purely aqueous supernatant is compatible with LC. On the other hand, the supernatant becomes strongly acidic using TCA ($1.4 < pH < 2.0$) or perchloric acid ($pH < 1.5$), and this may challenge the chemical stability of certain analytes. The acidic precipitants denature the plasma proteins and are believed to form insoluble salts with the positively charged amino groups of the plasma proteins at pH values below the isoelectric point, and this causes precipitation. Metal ions and the addition of salt can also be used for precipitation of plasma proteins, as shown in Table 6.1, but these are in use less often.

PPT is highly efficient for removal of plasma proteins in plasma and serum samples, and, as illustrated in Table 6.1, several approaches give a supernatant essentially free of proteins. PPT is not used for urine samples, as the content of proteins in urine is very low. In some cases, target analytes may be co-precipitated to some degree with the proteins, but the general experience with PPT is that the supernatant contains close to 100% of the analyte. The major advantages of PPT can be summarized to:

- Rapid procedure
- Requires minimal equipment
- Method development is relatively simple.

PPT is definitely a rapid procedure, and typically it can be completed within 10 minutes, including the final centrifugation (a single sample precipitated in centrifuge tubes). The equipment required is relatively simple, and it includes low-cost and disposable centrifuge tubes and a centrifuge. Ninety-six-well plates for PPT are more expensive, but still the cost per sample is low. The scientific literature contains a large number of published methods where PPT has been used, and often the bioanalytical scientist can find published methods ready for use. If some method development is required, this is relatively simple due to the simple nature of the technique. The principal optimization parameters include the type of precipitant, the amount of precipitant, and the centrifugation and filtration.

The major disadvantages of PPT can be summarized to:

- Limited sample clean-up
- No pre-concentration of the target analyte.

PPT can be very efficient in the removal of proteins, as discussed here, but a complex mixture of other endogenous and exogenous components will remain in the supernatant. Also, the plasma sample is actually diluted with the addition of the precipitant, and this will lower the concentration of the target analyte in the supernatant as compared to the original plasma sample.

6.3.3   Protein Precipitation in Modern High-Throughput Laboratories

PPT is typically performed in small centrifuge tubes, as discussed in this chapter. However, in modern bioanalytical laboratories, where many samples have to be analyzed every day, PPT can be performed in *multiwell plates*. In most cases, the multiwell plate is a *96-well plate*. This enables 96 samples to be protein-precipitated simultaneously. Ninety-six-well plates for PPT are commercially available from several manufacturers, and the principle is illustrated in Figure 6.2.

The 96-well system comprises two different plates, namely, a 96-well PPT plate (top) and a 96-well collection plate (bottom). The PPT plate contains filters in the bottom of each well, and the pore size of these filters is typically 0.2 μm. First, three sample volumes of acetonitrile (precipitant) are pipetted into the PPT plate. For example, 60 μl of acetonitrile

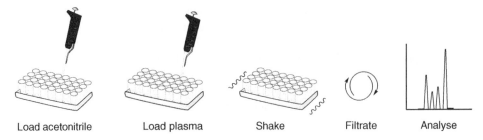

Figure 6.2   *96-well protein precipitation*

is used to precipitate proteins in 20 μl of plasma. Second, the plasma sample is pipetted into the PPT plate. Then, the PPT plate is sealed to avoid leaks, and the 96-well plate is shaken for 1–3 minutes. Finally, the samples are filtered through the 0.2 μm filters in the bottom of the PPT plate, and into a 96-well collection plate by vacuum, positive pressure, or centrifugation. The collection plate is located below the PPT plate. Thus, each well in the collection plate contains a clear liquid (filtrate) from a protein-precipitated sample. PPT in 96-well plates can typically be performed within 20 minutes; all steps are included and can be fully automated by laboratory robotics.

PPT can remove close to 100% of plasma proteins, as illustrated in Table 6.1. However, many other matrix components will remain in the supernatant because they are not precipitated. In recent years, substantial focus has been directed toward phospholipids. Phospholipids are present in human plasma and serum samples, and they are not removed efficiently by PPT. Phospholipids are known to cause ion suppression in LC-MS, which can affect seriously the reliability of quantitative measurements (discussed in Chapter 7). Therefore, 96-well phospholipid removal plates have recently been developed and commercialized, and they can be used in combination with PPT to remove both proteins and phospholipids from plasma and serum samples.

6.4   Liquid–Liquid Extraction

6.4.1   Basic Principles of Liquid–Liquid Extraction

LLE is intended to transfer target analytes from a biological fluid to an organic solvent. At the same time, most matrix components are intended to remain in the biological fluid, and this results in sample clean-up. LLE is most applicable to compounds of low polarity. The principle of LLE is illustrated in Figure 6.3.

The biological fluid (*sample*) is pipetted into a small centrifuge tube. Biological fluids like plasma, serum, whole blood, and urine are all aqueous. Next, pH in the sample is adjusted to ensure efficient extraction of the analyte, and this is accomplished by pipetting a buffer solution, or an acidic or basic solution, into the sample. Finally, an organic solvent (*extraction solvent*) is added to the tube. The extraction solvent should be immiscible with water, and it will form a two-phase system with the sample. Then, the tube is shaken for a certain time to mix the sample and the extraction solvent, and during this mixing the

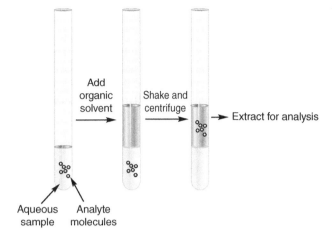

Add
organic
solvent

Shake and
centrifuge

→ Extract for analysis

Aqueous   Analyte
sample   molecules

Figure 6.3   *Principle of liquid–liquid extraction*

analyte molecules are transferred from the aqueous sample to the extraction solvent. The tube is typically shaken between 1 and 10 minutes. After this, the tube is allowed to stand for the two phases to separate. With biological fluids, this phase separation can be difficult and may take some time, but it is promoted by centrifugation of the mixture. Therefore, centrifugation is often performed for 1–3 minutes after extraction. After centrifugation, the extraction solvent can be collected as the *extract* (containing the analyte), and basically it is ready for the final analytical measurement. A typical example of an LLE procedure is illustrated in Box 6.2.

Box 6.2 Determination of 3,4-Methylenedioxymetamphetamine (MDMA) in Whole Blood Based on Liquid–Liquid Extraction

Background
MDMA (3,4-methylenedioxymetamphetamine) is a drug of abuse, and this substance is frequently determined in whole blood samples by forensic toxicology laboratories. The structure of MDMA is illustrated here:

HN

MDMA is a basic substance. The pK_a-value is 10.1, and the log P-value is 1.86. In the procedure presented here, MDMA is extracted from whole blood from drug abusers by LLE, and the final analytical measurement (not discussed here) is by LC-MS.

Procedure

Pipette 100 μl of whole blood into a centrifuge tube. Pipette 1 ml of 0.01 M NaOH (containing internal standard) and 1.5 ml of ethyl acetate into the centrifuge tube. Mix the contents on a vortexer for 30 seconds. Shake the centrifuge tube for 10 minutes mechanically. Centrifuge for 10 minutes at 4500 rpm. Collect the organic phase for the final analytical measurement.

Comments

Sodium hydroxide is added to adjust pH, and ethyl acetate is used as the extraction solvent.

6.4.2   Fundamentals of LLE

To fully understand LLE and the optimization of LLE procedures, looking into the fundamentals is highly important. This has been discussed briefly in Chapter 2, but here the discussion will be extended. From a fundamental point of view, LLE can be illustrated in the following way:

$$A_{Aqueous} \leftrightarrow A_{Organic} \tag{6.1}$$

where $A_{Aqueous}$ represents the analyte present in the sample (aqueous); and $A_{Organic}$ represents the analyte present in the extraction solvent (organic). The analyte is distributed between the two phases during LLE, and this distribution can be described mathematically by the *partition ratio* (K_D):

$$K_D = \frac{[A_{Organic}]}{[A_{Aqueous}]} \tag{6.2}$$

where $[A_{Organic}]$ is the concentration of the analyte in the extraction solvent; and $[A_{Aqueous}]$ is the concentration of the analyte in the sample at equilibrium. Equilibrium is normally established at the end of the extraction. K_D is constant for a given analyte under given conditions in the sample and in the extraction solvent. The higher the partition ratio, the more efficiently the analyte is extracted into the extraction solvent. Normally, bioanalytical chemists try to get as high extraction recoveries as possible during LLE method development, and this implies adjusting the extraction conditions to get as high a partition ratio as possible, according to Equation 6.2. For neutral analytes, Equation 6.2 describes the situation well. For these analytes, extraction recoveries are maximized by optimizing the type of extraction solvent. This will be discussed in more detail in Section 6.4.3. For neutral analytes, the pH value in the sample does not affect the extraction.

However, many analytes (pharmaceuticals) in the bioanalytical laboratory are either acidic or basic compounds, and for these the situation becomes a little more complicated because they tend to dissociate in aqueous solution. For an acidic analyte termed HA, dissociation in an aqueous sample occurs in the following way:

$$HA \leftrightarrow H^+ + A^- \tag{6.3}$$

The *dissociation constant* (K_a) for an acidic analyte can be expressed by the following equation:

$$K_a = \frac{[H^+] \cdot [A^-]}{[HA]} \tag{6.4}$$

For acidic analytes, the degree of dissociation (or ionization) is strongly dependent on pH. Equations 6.3 and 6.4 imply that the ionization of acidic analytes increases with decreasing $[H^+]$ and increasing pH. For an acidic analyte, the *distribution ratio* (D) can be defined according to the following equation:

$$D = \frac{[HA_{Organic,\,total}]}{[HA_{Aqueous,\,total}]} \tag{6.5}$$

where $[HA_{Organic,\,total}]$ is the total concentration of the analyte in the organic phase; and $[HA_{Aqueous,\,total}]$ is the total concentration of the analyte in the aqueous phase. The key point is that only the uncharged fraction of an acidic analyte can be extracted from the aqueous sample and into the organic solvent, whereas the charged species will remain in the aqueous sample. Thus, the following equations can be written for $[HA_{Organic,\,total}]$ and $[HA_{Aqueous,\,total}]$, respectively:

$$[HA_{Organic,\,total}] = [HA_{Organic}] \tag{6.6}$$

$$[HA_{Aqueous,\,total}] = [HA_{Aqueous}] + [A^-_{Aqueous}] \tag{6.7}$$

Substitution of Equations 6.6 and 6.7 into Equation 6.5 provides:

$$D = \frac{[HA_{Organic}]}{[HA_{Aqueous}] + [A^-_{Aqueous}]} \tag{6.8}$$

By substitution of Equations 6.2 and 6.4 into Equation 6.8, one gets:

$$D = K_D \cdot \frac{[H^+]}{[H^+] + K_a} \tag{6.9}$$

In a similar way, the following equation can be obtained for extraction of a basic analyte A, which partly dissociates (ionizes) in the aqueous sample:

$$A + H^+ \leftrightarrow AH^+ \tag{6.10}$$

$$D = K_D \cdot \frac{K_a}{[H^+] + K_a} \tag{6.11}$$

K_a is the dissociation constant for the corresponding acid AH^+. Also, for basic analytes, the degree of dissociation (or ionization) is strongly dependent on pH. Equation 6.10 implies that the ionization of basic analytes decreases with decreasing $[H^+]$ and increasing pH.

Equations 6.9 and 6.11 imply that distribution ratios for acidic and basic analytes are strongly pH dependent. This is illustrated in Figures 6.4 and 6.5. Figure 6.4 shows how the distribution ratio between 1-octanol and water varies as a function of pH for the acidic

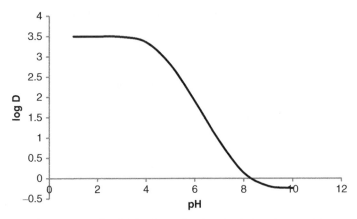

Figure 6.4   *Distribution ratio (log D) for ibuprofen (acidic drug substance) between 1-octanol and aqueous solution as function of pH*

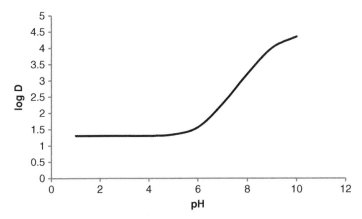

Figure 6.5   *Distribution ratio (log D) for amitriptyline (basic drug substance) between 1-octanol and aqueous solution as function of pH*

drug substance ibuprofen. Ibuprofen has a pK_a-value of 4.85. From this figure, it is clear that the highest distribution into the organic phase is obtained from acidic solution at pH values well below the pK_a-value of the compound. Note that the y-axis is logarithmic, and the distribution ratio increases by more than three orders of magnitude from alkaline to acidic solution. Mathematically, this behavior follows directly Equation 6.9. However, it is also consistent with general rules in chemistry. In alkaline solution, acidic analytes are charged. In their charged form, the water solubility is relatively good, and the solubility in an organic solvent is relatively poor. Thus, the analyte molecules will principally remain in the aqueous sample. On the other hand, in acidic solution, acidic analytes are uncharged. In their uncharged form, the water solubility is relatively poor, and the solubility in an organic solvent is relatively good. Thus, the analyte molecules will principally remain in the organic solvent. The practical consequence of this is that samples have to be acidified for efficient LLE of acidic analytes.

Figure 6.5 shows how the distribution ratio between 1-octanol and water varies as a function of pH for the basic drug substance amitriptyline. Amitriptyline has a pK_a-value of 9.76. From this figure, it is clear that the highest distribution into the organic phase is obtained from alkaline solution at pH values well above the pK_a-value of the compound. The distribution ratio increases by more than three orders of magnitude from acidic to alkaline solution. Mathematically, this behavior follows directly Equation 6.11. However, it is also consistent with general rules in chemistry. In alkaline solution, basic analytes are uncharged. In their uncharged form, the water solubility is relatively poor, and the solubility in an organic solvent is relatively good. Thus, the analyte molecules will principally remain in the organic solvent. On the other hand, in acidic solution, basic analytes are charged. In their charged form, the water solubility is relatively good, and the solubility in an organic solvent is relatively poor. Thus, the analyte molecules will principally remain in the aqueous sample. The practical consequence of this is that samples have to be made alkaline for efficient LLE of basic analytes.

From Equations 6.9 and 6.11, LLE of acidic and basic analytes is strongly dependent on pH, as discussed in this chapter. However, the partition ratio also plays a major role, and therefore the chemical composition or nature of the extraction solvent also is important during method optimization for LLE of acidic and basic analytes.

6.4.3 Solvents for LLE

To obtain high partition ratios and high extraction recoveries, the bioanalytical chemist has to select a strong or good solvent for the particular analyte. But how can we know or predict this? When analyte molecules are extracted and dissolved into an organic solvent in LLE, favorable interactions take place between the analyte molecules and the solvent molecules. In LLE, these molecular interactions can be of different types. For a given analyte, the type of potential interactions can be predicted from the molecular structure, and the solvent should be selected accordingly to facilitate these particular interactions. The predominant molecular interactions in LLE are the following:

- Hydrophobic interactions
- Dispersion interactions
- Dipole interactions
- Hydrogen bonding interactions.

The different molecular interactions are listed in order of increasing strength. Thus, hydrophobic interactions are weak, whereas hydrogen bonding interactions are strong. *Hydrophobic interactions* are nonpolar attractive interactions based on Van der Waals forces. Hydrophobic interactions act between portions of different molecules that are in close contact. Hydrophobic interactions principally occur between hydrocarbon moieties and related nonpolar molecular elements as shown in Figure 6.6a. The hydrophobic interactions are due to minor distortion of electron distributions, causing small and temporary dipole moments. Dissolution of an aliphatic hydrocarbon in hexane is a typical example where the dissolution is based on hydrophobic interactions.

Dispersion interactions, or induced dipole interactions, are weak attractive interactions between a relatively nonpolar electron-rich molecule and a polar (or charged) molecule as shown in Figure 6.6b. The polar molecule causes an electron density change in the nonpolar

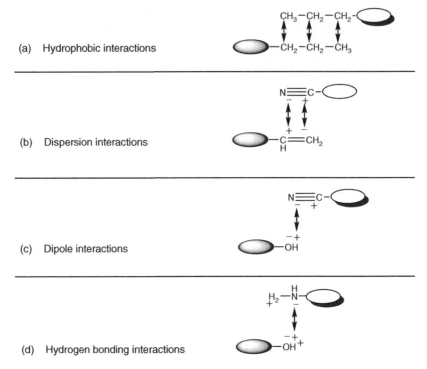

(a)   Hydrophobic interactions

(b)   Dispersion interactions

(c)   Dipole interactions

(d)   Hydrogen bonding interactions

Figure 6.6   *Overview of molecular interactions in liquid–liquid extraction*

molecule by making it slightly polar, or in other words, the polar molecule induce a dipole moment in the nonpolar and electron-rich molecule. Typical examples of nonpolar and electron-rich solvents which are prone to dispersion interactions are the aromatic solvents toluene and p-xylene. Thus, dissolution of polar substances in toluene is largely based on dispersion interactions. *Dipole interactions* are attractive electrostatic forces between two molecules with permanent dipole moment. Thus, the positive end of one molecule is attracted by the negative end of another molecule. This is illustrated in Figure 6.6c with an alcohol and a nitrile as example. Both components have high permanent dipole moments, and dissolution is based on dipole interactions. *Hydrogen bonding interactions* are another type of dipole interactions, where interaction occurs between the hydrogen atom that is part of a polar bond (*hydrogen bonding donor*) and an electronegative atom with a lone pair of electrons such as O and N (*hydrogen bonding acceptor*). This is illustrated in Figure 6.6d. Dissolution of a carboxylic acid in diethyl ether is a typical example where the dissolution is based on hydrogen bonding interactions. The acidic proton on the carboxylic acid serves as hydrogen bonding donor, whereas the lone pair electrons on oxygen in diethyl ether serve as a hydrogen bonding acceptor.

The knowledge about molecular interactions can be used to predict the best type of solvent for a specific LLE application. The selection of solvent should be based on the molecular structure of the analyte. Thus, for very nonpolar analytes, which are hydrocarbon like substances with no or very little functional groups, the principal interaction between

Table 6.2 *Frequently used liquid–liquid extraction solvents and their physiochemical properties*

Solvent	Polarity index	Boiling point (°C)	Water solubility (% w/w)	Density (g/ml)
Nonpolar solvents				
n-Pentane	0.1	35	0.004	0.65
Cyclohexane	0.2	81	0.008	0.79
n-Heptane	0.1	99	0.005	0.70
Aromatic solvents				
Toluene	2.4	111	0.03	0.87
p-Xylene	2.5	140	0.01	0.87
Dipolar solvents				
Dichloromethane	3.1	40	0.4	1.25
Ethyl acetate	4.4	74	4	0.90
Hydrogen bond donor solvents				
Chloroform[a]	4.1	61	0.2	1.50
Hydrogen bond acceptor solvents				
Methyl-tertbutyl ether	2.5	55	5	0.74
Diethyl ether	2.9	33	4	0.73

[a] Should be avoided if possible due to toxicity.

the analyte and the solvent has to be hydrophobic interactions. For this type of application, the best solvents will be very nonpolar solvents. Typical extraction solvents are aliphatic hydrocarbons, and n-pentane, n-heptane, and cyclohexane are frequently used as shown in Table 6.2.

For analytes with low to medium polarity, or for analytes with aromatic character, aromatic solvents like toluene or p-xylene are efficient. With toluene and p-xylene, dispersion interactions can take place. For even more polar analytes, a more polar solvent should be selected, but still not too polar because the solvent should be immiscible with water to form a two-phase system. For polar analytes, solvents with high dipole moments are very efficient, where dissolution by dipole interactions take place. Examples of popular solvents with good dipole properties are dichloromethane and ethyl acetate. For extraction of basic analytes, which are normally amines (containing nitrogen with lone pair electrons), solvents with hydrogen bonding donor properties are preferred. Chloroform is a solvent with strong hydrogen bonding donor properties, but several other solvents can be used as alternatives. For acidic analytes, which are characterized by their high hydrogen bonding donor properties, hydrogen bonding acceptor solvents should preferably be used. Typical examples of hydrogen bonding acceptor solvents are diethyl ether and methyl-tertbutyl ether as illustrated in Table 6.2.

As discussed above, the hydrogen bonding donor properties, the hydrogen bonding acceptor properties and the dipole interaction properties are important parameters in describing the dissolution power of a solvent. These parameters can be measured and quantified, and can be used to classify different solvents. A popular way to express these parameters are by the *Kamlet and Taft solvatochromic parameters* α, β, and π^* as illustrated in Table 6.3.

Table 6.3   *Kamlet and Taft solvatochromic parameters (α, β, and π^*) for selected solvents*

Solvent	α	β	π^*
Nonpolar solvents			
Cyclohexane	0.00	0.00	0.00
Aromatic solvents			
Toluene	0.00	0.11	0.54
Dipolar solvents			
Dichloromethane	0.13	0.10	0.82
Ethyl acetate	0.00	0.45	0.55
Hydrogen bond donor solvents			
Chloroform	0.20	0.10	0.58
Hydrogen bond acceptor solvents			
Diethyl ether	0.00	0.47	0.27

Values for α, β, and π^* can be found in the literature for many organic solvents. The α-value is a measure of the *hydrogen bonding donor property*. A high value for α indicate a strong tendency for the solvent to interact as a hydrogen bonding donor. Solvents with high α-values are especially good solvents for basic substances, and include chloroform and dichloromethane as frequently used LLE solvents. In general, small alcohols, phenols, and carboxylic acids have very high α-values, but are not useable for LLE due to unfavorably high water solubility. The β-value is a measure of the *hydrogen bonding acceptor property*. A high value for β indicates a strong tendency for the solvent to interact as a hydrogen bonding acceptor. Solvents with high β-values are especially good solvents for acidic substances. As illustrated in Table 6.3, ethers have high values for β, but also ethyl acetate is a relatively strong hydrogen bonding acceptor. Small amines and alcohols also have high β-values, but are generally not used in LLE due to unfavorably high water solubility. The π^*-value is a measure of the *dipolarity–polarizability* of a solvent, and give an indication of the tendency of the solvent to interact through dispersion interactions and dipole interactions. Solvents with high π^*-values are especially good solvents for polar substances. Dichloromethane and ethyl acetate are popular solvents with high π^*-values, as illustrated in Table 6.3. Also note from this table that many solvents have mixed properties. Ethyl acetate as an example is a strong solvent due to both hydrogen bonding acceptor properties and due to high dipolarity–polarizability.

As discussed here, polar analytes have to be extracted with relatively polar organic solvents, and nonpolar analytes have to be extracted with the more nonpolar solvents. The polarity of solvents can actually be quantified by the *polarity index* (P) as illustrated in Table 6.2. A solvent that can participate in strong molecular interactions, like hydrogen bonding interactions and dipole interactions, is a relatively polar solvent. For a polar solvent, the sum of α, β, and π^* is high, and the solvent has a high value for the polarity index. For the most nonpolar aliphatic hydrocarbons, like n-heptane, the polarity index is very close to 0. As the polarity increases, the polarity index also increases, and water has a polarity index of 10.2. Solvents with polarity index below 4–5 are normally immiscible with water and form a two-phase system with aqueous samples. These solvents can therefore be used in LLE. Generally, using solvents with very low polarity index, less matrix components are extracted. Solvents with a polarity index above 4–5 are totally miscible with water and cannot be used for LLE.

The choice of solvent for LLE will not solely be determined by the analyte solubility and distribution. Other properties of the solvents are also important for the final selection of LLE solvent. Thus, the *density* and *viscosity* are important properties for practical work. Solvents with low viscosity are preferred because they more easily brake up into small droplets and mix with the aqueous sample during extraction, and this is beneficial for the transfer of analyte into the organic phase. Solvents with a lower density than water will float on top of the aqueous sample after extraction. This is an advantage if the organic extract is to be collected by a pipette after extraction from a small centrifuge tube. Examples of solvents less dense than water are n-pentane and ethyl acetate. If a solvent more dense that water is used, it will be recovered below the aqueous sample after LLE. Examples of solvents with higher density than water are chloroform and dichloromethane. The volatility of the solvent and the solubility in water are also crucial. Any evaporation of solvent after extraction is much faster with a solvent with low boiling point, and volatile solvents are normally preferred if they have to be evaporated after extraction. Finally, for the solvent selected for extraction, the solubility in water should be low in order to avoid that substantial amount of the extracting solvent is dissolved into the aqueous sample. As seen in Table 6.2, the water solubility of solvents can be quite different; the nonpolar solvent n-heptane is practically insoluble in water, whereas ethyl acetate dissolves into water in substantial amounts.

6.4.4   Extraction Recovery in LLE

Equations 6.9 and 6.11 can be used to calculate the distribution ratio for LLE applications. These equations also demonstrate that the distribution ratio is very close to the partition ratio if the extraction is carried out at optimal pH (low pH for acids and high pH for bases). Under optimal pH conditions, the partition ratio can be used to calculate the expected recovery from LLE. Suppose that analyte A is extracted under optimal pH conditions from a certain volume of sample ($V_{Aqueous}$) and into a certain volume of solvent ($V_{Organic}$) with the partition ratio K_D and with the recovery R. Initially, m moles of A were present in the sample. After the LLE has been finished, the extraction solvent contains equal to $R \cdot m$ moles of A, whereas the sample contains equal to $(100 - R) \cdot m$ moles of A. This can be inserted into the equation for the partition ratio:

$$K_D = \frac{[A_{Organic}]}{[A_{Aqueous}]} = \frac{\dfrac{R \cdot m}{V_{Organic}}}{\dfrac{(100 - R) \cdot m}{V_{Aqueous}}} \tag{6.12}$$

Rearrangement of Equation 6.12 gives the following equation:

$$R = \frac{K_D \cdot V_{Organic}}{K_D \cdot V_{Organic} + V_{Aqueous}} \cdot 100\% \tag{6.13}$$

Equation 6.13 show that recoveries in LLE can be increased by (i) increasing the partition ratio, (ii) increasing the volume of extraction solvent, or (iii) decreasing the volume of sample. This is further illustrated by calculations in Box 6.3.

Box 6.3 Theoretical Recoveries with Different Partition Ratios and Volumes of Extraction Solvent

One milliliter of plasma is extracted with 1.0 ml organic solvent, and the partition ratio for the analyte in this particular extraction system is 10. The recovery is calculated based on the following equation:

$$R = \frac{10 \cdot 1.0}{10 \cdot 1.0 + 1.0} \cdot 100\% = 91\%$$

To improve the recovery, the volume of extraction solvent is increased to 5.0 ml. Now, the recovery is calculated to:

$$R = \frac{10 \cdot 5.0}{10 \cdot 5.0 + 1.0} \cdot 100\% = 98\%$$

Clearly, the recovery in LLE can be increased by increasing the volume of extraction solvent. Alternatively, one can try to optimize the extraction solvent to increase the partition ratio. Suppose that the extraction is now carried out with a solvent, providing a partition ratio of 30. The recovery is calculated to:

$$R = \frac{30 \cdot 1.0}{30 \cdot 1.0 + 1.0} \cdot 100\% = 97\%$$

Clearly, one can also increase the recovery in LLE by increasing the partition ratio by proper selection of extraction solvent. If both approaches are combined, increasing the volume of extraction solvent to 5.0 ml and increasing the partition ratio to 30, the theoretical recovery can be calculated:

$$R = \frac{30 \cdot 5.0}{30 \cdot 5.0 + 1.0} \cdot 100\% = 99\%$$

In the latter case, the extraction is essentially quantitative or exhaustive, which means that the total amount of analyte is transferred from the sample and into the organic phase.

From Equation 6.13, the expected recovery from any LLE of a certain analyte can be calculated when the volumes of sample and extraction solvent have been decided, provided that the partition ratio is known. But how can we get information about the partition ratio? Partition ratios vary from compound to compound, and they are very dependent on the experimental conditions. Especially, partition ratios are very dependent on the type of solvent used for LLE, as discussed in this chapter, but they also vary with temperature and with the presence of matrix components in the aqueous sample. Therefore, partition ratios cannot be found in general in the literature. Nevertheless, some valuable information can be obtained from the literature from tables with log P-values. As discussed earlier in this book (Chapter 2), the log P-value for a certain compound gives the logarithmic value for the partition ratio between 1-octanol and water. Theoretically, recoveries using 1-octanol as the extraction solvent can be calculated. However, 1-octanol is not used to any extent as extraction solvent in LLE, among others, because it has a high boiling point and is incompatible with most analytical instruments. However, a high log P-value indicates that the compound is relatively nonpolar, and the nonpolar analytes are the easiest to extract by LLE because they have high partition ratios in general. As a rule of thumb, compounds with log P-values below 0 are very polar and difficult to extract with any solvent in LLE. Compounds with log P between 0 and 1 are relatively polar and are not the best candidates for LLE. Compounds with log P-values above 1 are more hydrophobic, and they are generally well extracted in LLE.

6.4.5 Practical Work

Based on the discussion above, the extraction solvent for a certain analyte should in principle be selected based on the molecular structure of the analyte. However, for practical work, several other factors also have to be considered. First, the solvent has to be immiscible with water. This requires a polarity index of 4–5 or less. Second, the solvent should be volatile, especially when the extract has to be evaporated and reconstituted in another liquid. Third, the solvent should preferably be less dense than water, to form the upper layer after phase separation. This is convenient when the extract is collected after LLE. Finally, the organic solvent should be acceptable for use in terms of safety and environmental aspects. For example, diethyl ether should not be used as it forms explosive peroxides, 1-octanol should not be used as it is nonvolatile, benzene should be avoided as it is carcinogenic, and chloroform should be avoided for toxic and environmental reasons.

Relatively often, LLE is accomplished with *binary solvent mixtures*. Thus, rather than using a single pure solvent, two solvents are mixed. By mixing solvents, the polarity can be modified and tuned to improve analyte recovery. Typical examples are the addition of alcohols, such as 1-propanol or 1-butanol, to relatively nonpolar solvents. One example of this is illustrated in Box 6.4.

As mentioned, pH in the sample is important for LLE of acidic and basic substances. Normally, it is recommended to use a pH value in the sample at least two units from the pKa-value of the analyte. Thus, acidic analytes should preferably be extracted from sample solutions acidified to a pH value of two units below the pKa-value of the acids. For basic substances, pH in the sample should be adjusted two units higher than the pKa-value of the basic substances. The bioanalytical chemist should be careful when using extreme pH values in the sample. Thus, under strongly acidic or alkaline conditions, certain analytes can be prone to degradation. Also, at extreme pH values, precipitation of matrix components can occur in biological fluids, and this can sacrifice the LLE process. A practical example is illustrated in Box 6.4.

Box 6.4 LLE of Citalopram from Serum

Background

Citalopram is an antidepressant drug, and the typical concentration range of this substance in serum is 10–200 ng/ml. The structure of citalopram is illustrated here:

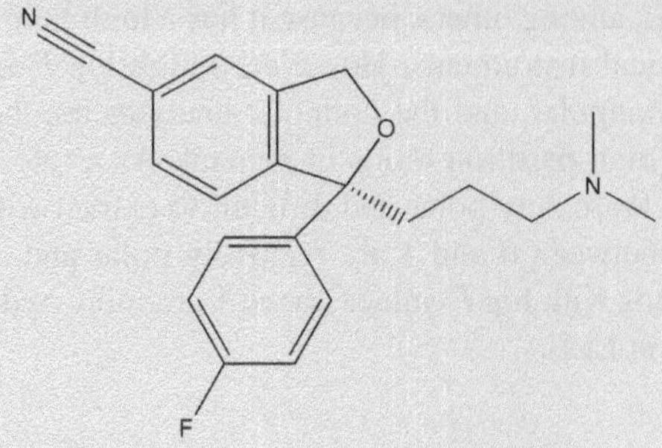

Citalopram is a basic substance with low water solubility. The pK_a-value is 9.8, and the log *P*-value is 3.76. In the procedure presented in this box, LLE is performed from serum samples from patients treated with citalopram, and the final analytical measurement (not discussed here) is by LC-MS.

Procedure

Pipette 100 μl of serum into a centrifuge tube. Pipette 50 μl of internal standard, 75 μl of saturated borate buffer (pH 10.5), and 1.2 ml of ethyl acetate/n-heptane (80:20, v/v) into the centrifuge tube. Mix the contents on a vortexer for 30 seconds. Shake the centrifuge tube for 10 minutes mechanically. Centrifuge for 10 minutes at 4500 rpm. Collect 750 μl of the organic phase, and evaporate to dryness at 40 °C. Reconstitute in 900 μl acetonitrile/5 mM ammonium acetate buffer pH 5.0 (25:75, v/v), and inject into LC/MS/MS.

Fundamental Questions to the Procedure—Step by Step

1. Why is the sample volume 100 μl?
2. Why is borate buffer pipetted into the sample?
3. Why is solvent volume 1.2 ml?
4. Why are ethyl acetate and n-heptane mixed?
5. Why is the centrifuge tube shaken for 10 minutes?
6. Why is the extract reconstituted in acetonitrile/ammonium acetate buffer?

Discussion of the Questions

1. Due to the high sensitivity of modern LC-MS instruments, only small sample volumes are required to measure drug substances. For safety and practical reasons,

small volumes are an advantage, but volumes below 50–100 µl can be difficult to pipette with high accuracy.

2. Borate buffer is added to the sample to adjust the pH to approximately 10.5. Under alkaline conditions, the target analyte is only partly ionized to a small extent, and extraction into the organic solvent is favored.

3. The solvent volume is large (1.2 ml) in order to maximize extraction recovery.

4. The perfect extraction solvent for citalopram should have significant hydrogen-bonding donor properties. Unfortunately, most solvents with these properties have some solubility in water, or they are unsuitable for LLE for other reasons. In this example, ethyl acetate and heptane were selected due to high volatility, and the solvent volume was increased to 1.2 ml to maximize extraction recovery. High volatility is an advantage because evaporation after LLE takes only a short time.

5. The centrifuge tube is shaken for 10 minutes to make sure that essentially all analyte molecules are transferred to the organic solvent.

6. The mixture of ethyl acetate and n-heptane cannot be injected into LC-MS. Therefore, the extract has to be evaporated and re-dissolved (reconstituted) in acetonitrile/ammonium acetate, which can be injected into LC-MS.

6.4.6   LLE with Back Extraction

LLE, as discussed so far, has been based on the fact that the analytes are transferred from the aqueous sample to an organic solvent (single extraction). Subsequently, the organic solvent or the organic extract is directed to the final chemical analysis. With very complicated samples containing many matrix components at high concentration, such a single extraction may not be sufficient as some of the matrix compounds also can be extracted into the organic solvent. In such cases, it can be an advantage to perform a second extraction step, called *back extraction*, into a new aqueous phase that is used for the final chemical analysis. Thus, the organic extract from the first extraction is extracted itself with a new aqueous phase in a second step. The advantages of back extraction are both improved cleanup of the sample and also that the final extract now is aqueous and compatible with LC. A typical scheme for extraction followed by back extraction is as follows: a basic drug substance is extracted from an alkaline aqueous sample to an organic solvent. Subsequently, the organic extract is collected into a new extraction tube, and an acidic aqueous solution is added as a second extraction fluid. Because basic compounds normally have high solubility in acidic solution due to protonation (ionization), the basic analyte is back-extracted from the organic solvent and into the acidic aqueous extract.

6.4.7   LLE in Modern High-Throughput Laboratories

LLE can be performed in the 96-well format. The major benefit of this is that 96 samples can be extracted simultaneously, and this of course increases the sample throughput dramatically. The most popular format of 96-well LLE is *supported liquid extraction* (SLE), as illustrated in Figure 6.7.

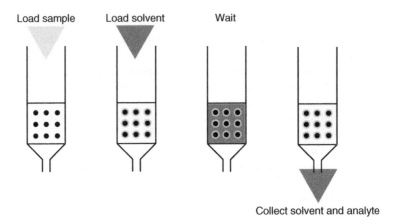

Figure 6.7   *Principle of supported liquid extraction*

The 96-well plate for SLE consists of 96 individual wells packed with a modified diatomaceous earth material as solid support. First, the aqueous sample is loaded into the well, and the sample is fully adsorbed as a thin layer on the surface of the support. Then, the extraction solvent, which should be immiscible with water, is loaded into the well. The solvent is typically retained in the well for a few minutes to accomplish the SLE process. The high surface area at the interface between the organic and aqueous phases maximizes extraction efficiency (analyte recovery) and minimizes the possibility of emulsion formation. The analytes are eluted from the well as the solvent passes through the well, and they are collected in a suitable collection plate. Automated operation of the SLE process is accomplished with a 96-sample vacuum-processing station, where vacuums at the well outlet suck the organic solvent through the solid support. The theory discussed in this chapter for LLE is also valid for SLE. The main advantage of SLE is that up to 96 samples can be extracted simultaneously. In addition, mixing, shaking, and centrifugation, which are steps required in traditional LLE, are avoided in SLE.

6.5   Solid-Phase Extraction

6.5.1   Fundamentals

SPE is frequently used for sample preparation of biological fluids like plasma, serum, and urine. The principle of SPE is illustrated in Figure 6.8. SPE uses a small extraction column (tube) filled with a stationary phase. The first step in the procedure includes *conditioning* the column by flushing certain liquids through the column. The purpose of the conditioning is to make the column ready for the application, and this will be discussed in more detail in Section 6.5.3. The second step is the sample *loading*, where the sample solution is flushed through the column. During sample loading, the analytes are retained in the column by different interactions with the stationary phase, whereas matrix components are flushed out of the column. In the third step, termed *washing*, one or more washing solutions are flushed through the column. The purpose of the washing is to flush and remove additional

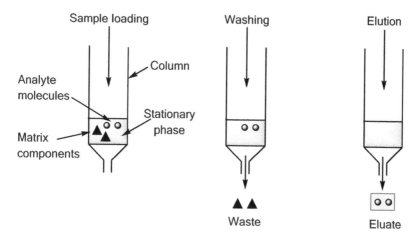

Figure 6.8   *Principle of solid-phase extraction*

matrix components, while the analytes are still retained in the column by strong interactions with the stationary phase. This step gives sample clean-up and is important to eliminate potential interferences in the final chemical analysis. In the final step, termed *elution*, a small volume of a certain liquid, which is called the *eluent*, is flushed through the column to break the interactions between the analytes and the stationary phase, and to release the analytes from the column. This final solution, which is called the *eluate*, is collected at the bottom of the column. The eluate contains the analyte, free of major matrix compounds from the sample, and the eluate is used for the final chemical analysis. SPE is normally a very efficient extraction method, and extraction recoveries are close to 100%.

6.5.2   The SPE Column

Figure 6.9 shows (a) a photo and (b) an illustration of an *SPE column*. The column (tube) itself is normally made of polypropylene. A certain mass of stationary-phase particles are packed into the bottom of the column, and the stationary-phase particles are held in place by two polyethylene filters. The size of the stationary-phase particles is typically in the range of 40–50 µm. Particles of this size allow a simple vacuum at the column outlet to obtain liquid flow through the system, and one can easily flush biological fluids with relatively high viscosity through the column.

A broad range of SPE columns is commercially available, with different masses and chemistries of the stationary phase. The typical SPE column for 1 ml samples contains 30–100 mg of stationary phase. As will be discussed in this chapter, a broad range of different stationary phases is available, but they are based on either *silica particles* or *organic polymeric particles*. This means that *functional groups* are the active part of the stationary phase, and these functional groups are attached to the surface of the particles in large numbers. Silica-based particles have for many years been very popular. Those are highly porous particles with an average pore size of 60 Å. Low-molecular drugs and other small molecules can diffuse into the pores, and the surface in the pore structures accounts for

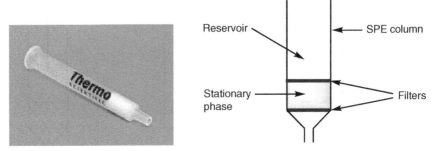

Figure 6.9 *(a) Photo and (b) illustration of a solid-phase extraction (SPE) column. A broad range of SPE columns are commercially available, with different masses and chemistries of the stationary phase*

approximately 98–99% of the total surface of the particle. Thus, low-molecular compounds can be retained very efficiently. In contrast, for macromolecules like proteins, their large size prevents them from entering the pore structures, and they are just exposed to 1–2% of the particle surface. In other words, macromolecules are just slightly retained in the column, they are easily flushed from the column, and in this way one can easily remove them by SPE. An SPE column containing a certain mass of stationary phase has a limited *capacity* for retaining analytes. Normally, the capacity is approximately 5% of the mass of the stationary phase. This means that a column packed with 100 mg stationary phase has a total capacity of 5 mg. As long as the mass of analyte and matrix components initially retained by the column is less than 5 mg in total, the SPE column is not *overloaded*. The bioanalytical chemist should always avoid overloading SPE columns, as this can result in lower and nonrepeatable extraction recoveries.

6.5.3   Conditioning

As mentioned in this chapter, the first step in the SPE procedure is *conditioning* of the SPE column. When SPE columns are obtained from a manufacturer, they are delivered in a dry state. In a dry state, the functional groups are inactive, and they have to be solvated before they can retain our analytes. *Solvation* is carried out by flushing the column with a polar organic solvent, and the typical solvent for this purpose is methanol. Figure 6.10 illustrates the effect of solvation. After solvation, the functional groups have been "raised," and they are ready to retain the analytes when the sample is loaded into the column. After flushing with methanol, the excess of methanol that is located in the *bed volume* (i.e., the volume between the particles) has to be removed. The reason for this is that methanol is a strong eluent in most SPE, and if methanol is present in the bed volume when the sample is applied, the analytes can be poorly retained and partly lost. Methanol is normally removed in the second part of the conditioning process by flushing with water or with aqueous buffers.

Conditioning is very important with silica-based SPE stationary phases, whereas for some organic polymeric-based phases, conditioning is actually not required. Conditioning also serves another purpose (i.e., to rinse the SPE column for impurities related to production), and this may be advantageous with all types of SPE stationary phases.

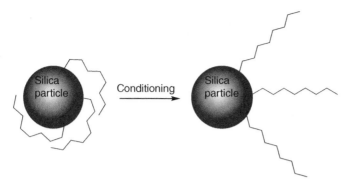

Figure 6.10   *Conditioning and solvation of a solid-phase extraction stationary phase*

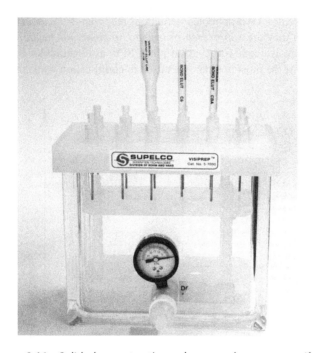

Figure 6.11   *Solid-phase extraction columns and vacuum manifold*

6.5.4   Equipment

SPE can be accomplished with relatively simple equipment, as illustrated in Figure 6.11, and the key elements are the SPE column and a *vacuum manifold*.

SPE columns are available from several manufacturers. Normally, a SPE column is used for only a single extraction, then it is discarded. In addition, a vacuum manifold is needed. The purpose of this is to create a vacuum at the column outlet that forces the liquids through the column. At the connection between the bottom of the column and the manifold, a valve is placed to turn "on" and "off" the vacuum within the column. The vacuum is turned on when a liquid is to pass the column, whereas the vacuum is turned "off" in between to avoid the column drying out. In the vacuum manifold, small vials are placed under each column to collect the final eluate from each extraction. In this way, several samples can be extracted simultaneously.

6.5.5   Reversed-Phase SPE

Reversed-phase SPE is used for extraction of relative nonpolar analytes from aqueous sample solutions. Extraction is based on *hydrophobic interactions* between the analyte and the stationary phase, and the principle is very similar to reversed-phase chromatography, as discussed earlier in Chapter 4. This means that retention increases with increasing hydrophobicity of the analyte. The interactions are based on *van der Waals forces*, primarily between hydrogen–carbon bonds in the stationary phase and hydrogen–carbon bonds in the analyte molecules. Hydrophobic interactions are *promoted in aqueous or polar environments*, whereas they are *suppressed in organic or nonpolar media*. Several different stationary phases are available for reversed-phase SPE, as illustrated in Figure 6.12, which summarizes some of the most popular ones used in combination with silica particles.

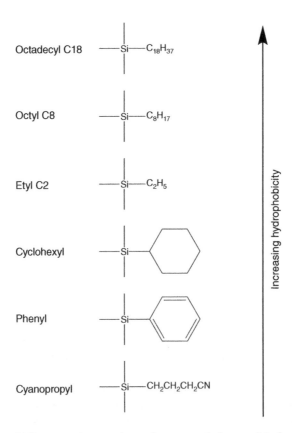

Figure 6.12   *Different stationary phases for reversed-phase solid-phase extraction*

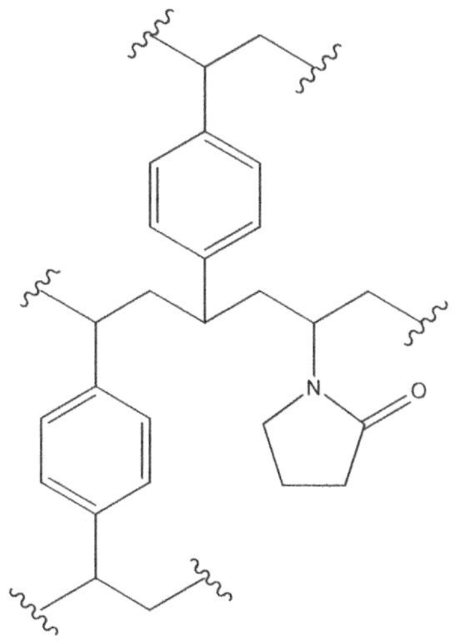

Figure 6.13   *Example of polymeric solid-phase extraction stationary phase*

The different stationary phases have different sizes of the hydrophobic group. C18-columns are very popular, and they are also the most hydrophobic reversed-phase columns available. Because of their strong hydrophobicity, C18-columns give strong retention to many compounds. This means that C18-columns are very general and can be used for many applications. However, because this column can extract a broad range of organic compounds, it does not give the best selectivity. In order to increase the selectivity, one can use some of the less hydrophobic phases like C8-columns, which are also very popular.

In addition to silica-based SPE columns, one can also use polymeric-based SPE columns for reversed-phase SPE. One example of a polymer used as the stationary phase in SPE is illustrated in Figure 6.13.

The stationary phase in Figure 6.13 is a co-polymer built from two different monomers: the hydrophobic monomer divinylbenzene and the more polar N-vinylpyrolidone. The divinylbenzene moieties give the stationary phase hydrophobicity, and these are responsible for hydrophobic interactions with the analytes. In contrast, the N-vinylpyrolidone moieties give the stationary phase polar functions that can retain more polar compounds. An important advantage of the polymer-based stationary phases is that they do not change chemical properties even if they dry out during the procedure, and the initial conditioning step may not be required.

As mentioned in this section, hydrophobic interactions are promoted in aqueous or polar environments, whereas they are suppressed in organic or nonpolar media. This means that a relatively nonpolar drug present in an aqueous biological fluid is an ideal case for reversed-phase SPE. The aqueous sample can be loaded directly into the SPE column. The nonpolar nature of the analyte results in strong retention by hydrophobic interactions with the stationary phase. The aqueous environment promotes the hydrophobic interactions between the analyte and the stationary phase. For acidic or basic analytes, retention can be even better if the compounds are present in the sample in their uncharged state. This means that, for basic analytes, alkaline conditions in the sample can improve the retention of the analyte on the SPE column. This can be accomplished by diluting the sample with an alkaline buffer.

During washing, pure water or other 100% aqueous solutions are normally used to remove matrix components from the SPE column. This will remove water-soluble compounds that have no affinity to the stationary phase, which means polar compounds and inorganic salts. By keeping the washing solution 100% aqueous, one can still have strong retention of the analytes based on hydrophobic interactions. In some cases, if the analytes are very well retained, aqueous washing solutions containing 10% or 20% methanol or other organic solvents can be used, but care should be taken not to start eluting the analytes from the column. Small amounts of methanol in the washing solution will remove more interference from the column and give a cleaner extract.

In the final elution of the analytes, 100% methanol or mixtures of methanol and water are typically used. Pure methanol gives a 100% organic medium in the column, and the hydrophobic interactions are totally suppressed. The key point during elution is that the hydrophobic interactions should be suppressed immediately in order to elute the analytes from the column in the lowest volume of eluent possible. The reason for using small volumes of eluent is to avoid dilution of the eluate. If the analytes are basic or acidic compounds, pH in the eluate can be adjusted to charge the analytes, as charging the analytes further reduces hydrophobic interactions. A typical procedure for reversed-phase SPE is discussed in Box 6.5.

Box 6.5 SPE of 3,4-Dihydroxymethamphetamine 3-Sulfate (DHMA 3-Sulfate) from Urine

Background

DHMA 3-sulfate (3,4-dihydroxymethamphetamine 3-sulfate) is a metabolite of MDMA. MDMA is a drug of abuse. The structure of DHMA 3-sulfate is illustrated here:

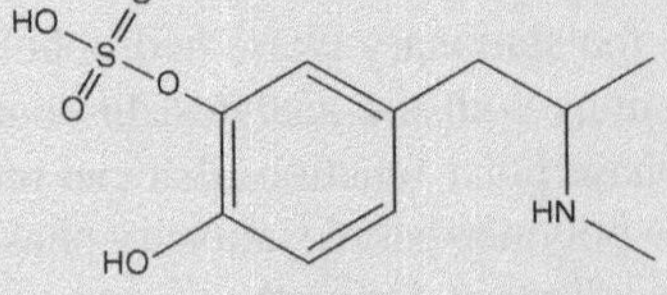

Procedure

Pipette 100 µl of urine sample into a vial. Add 10 µl of internal standard and 1 ml of water to the sample. A 500 mg C18 cartridge is conditioned with 1 ml of methanol and 1 ml of water. Load the prepared sample, and wash with 0.5 ml of water. Elute with 1 ml of methanol, and collect the eluate for LC-MS.

Fundamental Questions to the Procedure—Step by Step

1. Why is the sample volume only 100 μl?
2. Why is internal standard added?
3. Why is water added?
4. Why is a C18 cartridge used?
5. Why is the cartridge conditioned with 1 ml of methanol?
6. Why is the cartridge conditioned with 1 ml of water?
7. Why is the cartridge washed with 0.5 ml of water?
8. Why is the cartridge eluted with 0.5 ml of methanol?

Discussion of the Questions

1. Due to the high sensitivity of modern LC-MS instruments, only small sample volumes are required to measure drug substances. For practical reasons, small volumes are an advantage, but volumes below 50–100 μl can be difficult to pipette with high accuracy.
2. Internal standard is added to the sample to improve the precision and accuracy of the method.
3. Water is added to increase the sample volume for practical reasons.
4. C18-columns are superior for isolation of analytes with hydrophobic moieties from aqueous samples.
5. Conditioning with methanol is required for solvation of the SPE column.
6. Conditioning with water is required to remove excess methanol from the bed volume in order to avoid loss of analyte during sample application.
7. The cartridge is washed with water to remove water-soluble matrix components from the column—this gives clean-up.
8. The cartridge is eluted with a small volume of a strong eluent (methanol)—this immediately releases the analyte from the column.

6.5.6   Secondary Interactions in Reversed-Phase SPE

With silica-based stationary phases, the functional groups are attached to the surface of the silica particles. However, between the functional groups, *residual silanol groups* are present, as illustrated in Figure 6.14.

Silanol groups are acidic, and in aqueous medium, they can retain analytes due to ionic interactions. These interactions are called *secondary interactions*. Secondary interactions can be advantageous, because analytes participating in both hydrophobic interactions and secondary interactions are strongly retained in the column. In such cases, stronger washing solutions can be used without losing the analyte, resulting in better sample clean-up. For example, pure acetonitrile can be used during washing, and this solvent will remove many both nonpolar and more polar matrix components, whereas the analyte is still retained by the secondary interaction. To elute analytes retained by secondary

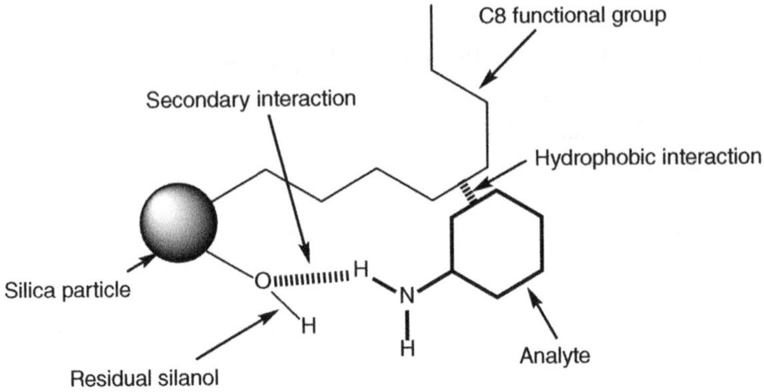

Figure 6.14   *Secondary interactions in solid-phase extraction*

interactions, methanol mixed with an acid is normally used. Methanol suppresses the hydrophobic interactions, whereas the acid reduces the ionization of the silanol groups and suppresses the secondary interaction, facilitating the analyte to be eluted from the column. An example of a reversed-phase SPE method involving secondary interactions is discussed in Box 6.6.

Box 6.6 SPE of Trimipramine from Serum

Background

Trimipramine is an antidepressant drug. Trimipramine is a basic substance ($pK_a = 9.42$) with a log P-value of 4.76. The structure of trimipramine is illustrated here:

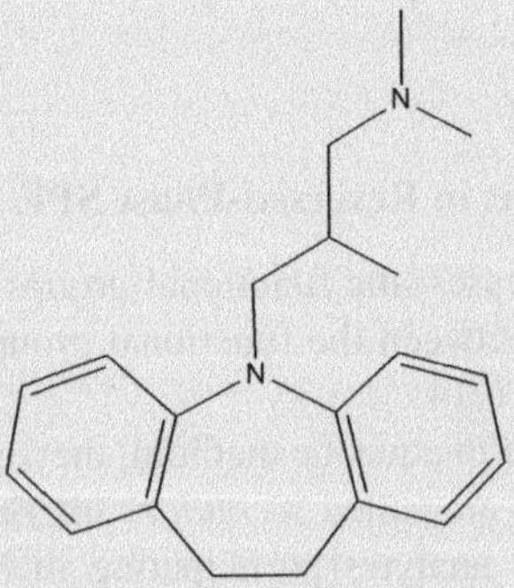

Procedure

A 500 mg C18 cartridge is conditioned with 2×1 ml of methanol and 2×1 ml of water. Load 1 ml serum sample, and wash with 2×1 ml of water. Elute with 0.5 ml of 10 mM acetic acid in methanol, and collect the eluate for LC-MS.

Fundamental Questions to the Procedure—Step by Step

1. Why is a C18 cartridge used?
2. Why is the cartridge conditioned with 2 × 1 ml of methanol?
3. Why is the cartridge conditioned with 2 × 1 ml of water?
4. Why is the cartridge washed with 2 × 1 ml of water?
5. Why is the cartridge eluted with 0.5 ml of 10 mM acetic acid in methanol?

Discussion of the Questions

1. C18-columns are superior for isolation of analytes with hydrophobic moieties from aqueous samples.
2. Conditioning with methanol is required for solvation of the SPE column.
3. Conditioning with water is required to remove excess methanol from the bed volume.
4. The cartridge is washed with water to remove water-soluble matrix components from the column—this gives clean-up.
5. The analyte may be retained also by secondary interactions between silanols and the tertiary amino group in the analyte, and elution with pure methanol is not efficient. With acetic acid in the eluent, secondary interactions are effectively suppressed because acetic acid suppresses the ionization of the silanol groups.

6.5.7   Ion Exchange SPE

Analytes containing ionic groups, like acids and bases, can be extracted from aqueous samples by *ion exchange SPE*. The principal type of interactions in ion exchange SPE is *ionic interactions* between ionic groups on the stationary phase and ionic groups on the analytes. Cationic analytes are positively charged, and they are extracted with *cation exchangers*, which are negatively charged. Many low-molecular drug substances are amines, and they can be extracted with cation exchangers. In a similar way, anionic analytes are negatively charged, like carboxylic acids (at neutral and high pH), and they can be extracted with *anion exchangers*, which are positively charged.

Figure 6.15 gives an overview of some ion exchangers that are used frequently for SPE. As illustrated, the stationary phases are characterized as *strong ion exchangers* or *weak ion exchangers*. A strong ion exchanger is ionized in the entire pH range, and it will act as ion exchanger regardless of pH in the SPE column. In contrast, a weak ion exchanger is ionized only in parts of the pH range, and weak ion exchangers can be turned "on" or "off" depending on pH in the SPE column.

Strong cation exchangers typically use *sulfonic acid* functionalities as the ionic group. The sulfonic acids have very low pK_a-values; this means that they are strongly acidic and therefore they are charged almost in the entire pH range. The strong cation exchangers work as ion exchangers in the entire pH range. Weak cation exchangers are often based on *carboxylic acid* functionalities as the ionic group. Carboxylic acids are weak acids, with pK_a-values around 4.5–5. This means that if pH in the SPE column is adjusted to above pH 3, the ion exchanger is gradually turned "on," and if pH is reduced well below pH 3,

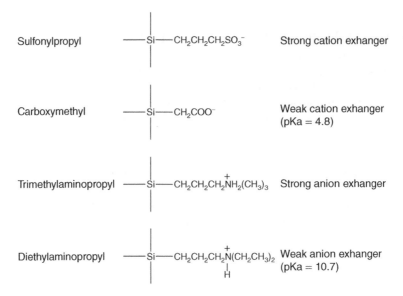

Figure 6.15   *Overview of some stationary phases for ion exchange solid-phase extraction*

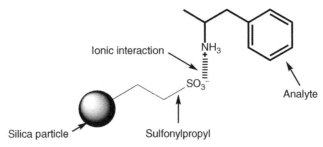

Figure 6.16   *Retention of amphetamine on a strong cation exchange solid-phase extraction column*

the ion exchanger is turned "off." As discussed later in this chapter, this can be utilized to control retention and subsequent elution of analytes in ion exchange SPE.

Anion exchangers can also be divided into strong and weak materials. Strong anion exchangers typically contain *quaternary ammonium* moieties as the functional group. Quaternary ammonium groups are ionized in the entire pH range. This means that strong anion exchangers are always turned "on." Weak anion exchangers normally contain a *secondary or tertiary amine* as the functional group, and these have pK_a-values typically in the vicinity of 10. Therefore, weak anion exchangers can be turned "off" by adjusting pH above 10 in the SPE column, whereas they can be turned "on" if pH is lowered well below pH 10. Figure 6.16 illustrates retention of amphetamine as an analyte on a strong cation exchange SPE column.

During sample loading in Figure 6.16, it is important that both the ion exchanger and the analyte are charged to promote ionic interactions and retention. Because a strong cation exchanger is used in this case, it is charged (turned "on") in the entire pH range. However, it is important to be careful with pH in the sample, because during loading, amphetamine should be in its charged form. Therefore, pH is adjusted well below the pK_a-value of amphetamine. The pK_a-value of amphetamine is about 10.1, and it is generally recommended to adjust pH at least 2 units below the pK_a-value for basic analytes to be sure that the compound is totally ionized. This means that pH in the sample related to Figure 6.16 should not exceed 8.0. If pH adjustment is required in the sample, one prefers to use buffers because this gives efficient control of pH. In general, the same holds for acidic analytes. They also have to be ionized during loading, and this requires a pH in the sample exceeding pK_a with at least 2 units.

In general, to ensure strong retention of the analyte during sample loading, it is also important to have control on the content of other ions (matrix ions) in the sample. The reason for this is that matrix ions at very high concentration, or with very high affinity for the ion exchanger, can compete with the analyte for the active sites on the stationary phase. If competition occurs, the analyte can be partly flushed through the column during loading and can be lost. This, in turn, will give low extraction recoveries. High content of matrix ions in the sample is especially challenging when extracting urine samples, which contain many different ions that can vary in concentration from sample to sample. In such cases, the sample can be diluted with pure water prior to extraction in order to reduce the concentration of matrix ions.

After loading, the SPE column is washed in order to remove as many matrix components as possible. Normally, water or preferably buffer solutions are used at this stage. One has to be careful with pH in order not to turn "off" the ionization of the analyte during washing. If the ionization of the analyte is partly turned "off" during washing, the analyte will be lost. This, in turn, will give low extraction recoveries and problems related to repeatability. One can also wash with methanol because methanol does not disturb the ionic interactions between the analyte and the stationary phase. Washing with methanol can be important to remove neutral matrix compounds with low polarity, as they can be retained in the column by hydrophobic interactions to the hydrocarbon chains located between the SPE particle and the ionic head group (see Figure 6.15).

In the final step, the analyte is eluted from the SPE column. In the example from Figure 6.16 with amphetamine, one can elute the analyte with an eluent with pH 12, which ensures that the charge on amphetamine is turned "off" and the ionic interactions are immediately broken. Addition of methanol to the eluent is probably an advantage as amphetamine also can have some hydrophobic interaction with the hydrocarbon chains of the ion exchanger. Methanol will also increase the solubility of amphetamine in the eluent. In general, elution in ion exchange SPE is based on turning "off" the ionic interactions. With strong ion exchangers, the ionic interactions are turned "off" by turning "off" the charge on the analyte. With weak ion exchangers, on the other hand, one can turn "off" the ionic interactions by turning "off" the ionization of the ion exchanger. In all of the cases, turning "off" the ionic interactions involves the selection of appropriate pH conditions in the eluent.

6.5.8   Mixed-Mode SPE

In *mixed-mode SPE*, both hydrophobic interactions and ionic interactions are used for analyte retention. An example is illustrated in Figure 6.17 with amphetamine as the analyte.

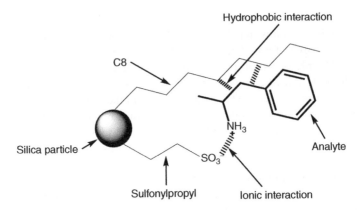

Figure 6.17   *Retention of amphetamine on a mixed-mode solid-phase extraction column*

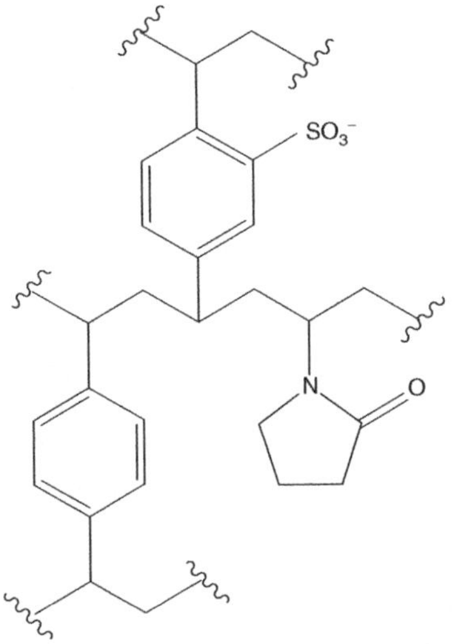

Figure 6.18   *Polymeric-based stationary phase for mixed-mode solid-phase extraction*

This particular silica-based sorbent contains C8-groups for hydrophobic interactions with the analyte, and strong cation exchanger groups for ionic interactions with the analyte. Polymeric-based SPE stationary phases can also be purchased, and one example is illustrated in Figure 6.18. In this case, sulfonate groups have been included in the polymer to give ionic interactions with cations.

It should be emphasized that mixed-mode SPE is not limited to the stationary phases shown in Figures 6.16 and 6.17, but a broad range of different mixed-mode materials are commercially available.

During sample loading, hydrophobic interactions are normally used as the primary source of retention. This is because hydrophobic interactions are less affected by variations in the sample matrix than ionic interactions. During washing in the next step, one first washes with a buffer that turns "on" the ionic interactions between the analyte and the stationary phase. pH in this solution should ionize both the stationary phase and the analyte. Washing with this buffer will also remove many polar matrix components. One can then wash the column with pure methanol because the analyte also is retained by ionic interactions, which are not broken by methanol. By washing with methanol, neutral matrix components of low polarity can be removed. This is the reason why mixed-mode SPE gives very clean extracts from biological fluids.

For elution of the analytes in mixed-mode SPE, one needs to turn "off" both the hydrophobic interactions and the ionic interactions. The hydrophobic interactions are suppressed by using methanol as one of the constituents in the eluent. The ionic interactions are suppressed by adding either a base or an acid to the eluent. For basic analytes, base is added to the eluent. High pH in the eluent will suppress the ionization of the analyte, and often methanol mixed with ammonia is used as eluent. A typical procedure for mixed-mode SPE is discussed in Box 6.7.

Box 6.7 SPE of Doping Agents (Anabolic Agents, β2-Agonists, Hormone Antagonists, Diuretics, Stimulants, Narcotics, Glucocorticoids, and β-Blockers) from Urine

Background
The analytes represent a broad range of chemical structures, but all substances can be retained by hydrophobic interactions. In addition, several of the analytes can be retained by ionic interactions.

Procedure
Prior to extraction, 1 ml of each urine sample is added to 100 μl of internal standard solution (mefruside 10 μg/ml and D_3-epitestosterone 1 μg/ml in methanol) and 1 ml of β-glucuronidase working solution (β-glucuronidase in phosphate buffer pH 6.2). After vortex mixing, each sample is incubated for 1 hour at 50 °C. After hydrolysis, 2 ml of 2% formic acid solution in water is added to each sample. Prior to loading, each sample is centrifuged.

A 60 mg mixed-mode SPE column (containing both hydrophobic moieties and cation exchange functionalities) is conditioned with 0.5 ml methanol and 0.5 ml of 2% formic acid solution. After this, the sample is loaded, and the column is washed with 1 ml of 2% formic acid in water, 1 ml of water, and 1 ml of 20% methanol in water. Elution is performed with 3 ml of 3% ammonium hydroxide in methanol–acetonitrile (50:50 v/v).

Fundamental Questions to the Procedure—Step by Step
1. Why is internal standard added?
2. Why is β-glucuronidase added?
3. Why is incubation performed for 1 hour at 50 °C?

4. Why is 2% formic acid solution in water added to each sample after incubation?
5. Why is the sample centrifuged prior to loading?
6. Why is mixed-mode SPE used?
7. Why is the cartridge conditioned with 0.5 ml of methanol?
8. Why is the cartridge conditioned with 0.5 ml of 2% formic acid solution?
9. Why is the cartridge washed with 1 ml of 2% formic acid in water?
10. Why is the cartridge washed with 1 ml of water?
11. Why is the cartridge washed with 1 ml of 20% methanol in water?
12. Why is the elution performed with 3 ml of 3% ammonium hydroxide in methanol–acetonitrile (50:50 v/v)?

Discussion of the Questions

1. Internal standard is added to improve the precision and accuracy of the method.
2. β-glucuronidase is an enzyme that is added to the sample to convert urinary glucuronides of the target analytes into the free form of the substances.
3. Elevated temperature and relatively long time (1 hour) are required to make sure that the enzymatic reaction is complete.
4. Acidic conditions in the sample are important to protonate basic analytes—in this way, they can interact through ionic interactions with the stationary phase.
5. Centrifugation is performed to remove particulate matter, as this can clog the SPE column.
6. Mixed mode is used to make sure that all substances of interest are extracted by hydrophobic interactions, ionic interactions, or both.
7. Conditioning with methanol is required for solvation of the SPE column.
8. Conditioning with 0.5 ml of 2% formic acid solution is required to remove excess methanol from the bed volume and to obtain acidic conditions in the bed volume.
9. Washing with 1 ml of 2% formic acid in water is performed to remove polar and acidic matrix components.
10. Washing with 1 ml water is performed to remove polar and acidic matrix components.
11. Washing with 1 ml of 20% methanol in water is performed to remove matrix components with weak retention, and this further improves the clean-up.
12. Methanol and acetonitrile are used to suppress hydrophobic interactions, and ammonium hydroxide is used to suppress ionic interactions.

6.5.9   Normal-Phase SPE

Normal-phase SPE can be used to isolate polar analytes present in non-aqueous sample solutions. Normal-phase SPE is much less in use in the bioanalytical laboratory than reversed-phase SPE, but it will be discussed briefly in this section to complete the introduction to SPE. In normal-phase SPE, the analytes are retained in the column by *polar interactions* with the stationary phase. The polar interactions are the same as discussed under normal-phase chromatography (Chapter 4), namely, *hydrogen bonding, dipole–dipole interactions, induced dipole–dipole interactions,* and *π–π interactions.*

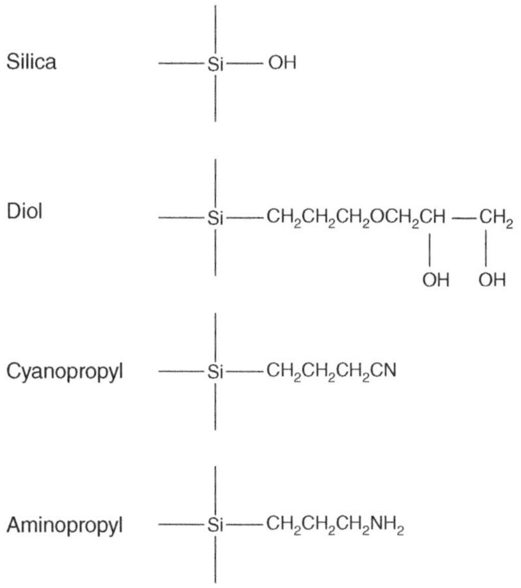

Figure 6.19 *Overview of some stationary phases for normal-phase solid-phase extraction*

For an analyte to give polar interactions, it should contain one or more polar groups in its structure. Thus, compounds containing hydroxyl groups, amines, carbonyl groups, aromatic rings, double bonds, or heteroatoms like oxygen, nitrogen, sulfur, or phosphorous can be extracted by normal-phase SPE. Figure 6.19 illustrates some sorbents frequently used in normal-phase SPE.

The typical functional groups are hydroxyl groups, amino groups, or cyanopropyl groups. Even pure silica particles, without any external groups attached to it, work as stationary phase for normal-phase SPE. In the latter case, the silanol groups on the surface of the silica particles serve as the functional group.

Normal-phase SPE frequently relies on hydrogen bonding between the analyte and the stationary phase. Hydrogen bonding is typical when hydrogen is bonded to electronegative atoms like oxygen or nitrogen. Especially, strong hydrogen bonding occurs between hydroxyl and amino groups. The polar interactions are turned "on" in nonpolar environments, whereas they are turned "off" in polar environments. This means that the sample should be dissolved in a nonpolar or low-polarity organic solvent during sample loading. On example of such a nonpolar solvent is n-hexane. For elution of the analyte, one has to change to a much more polar solvent that itself can hydrogen bond to the stationary phase, and that can elute the analyte from the column. One example of a polar solvent that can be used to suppress hydrogen bonding interactions is methanol. Methanol can thus be used for elution of the analytes.

6.5.10 SPE in Modern High-Throughput Laboratories

In modern high-throughput laboratories, SPE is often performed in 96-well format, as illustrated in Figure 6.20. The great advantage of SPE in 96-well format is that 96 samples can

Figure 6.20   *Photo of a solid-phase extraction 96-well plate*

be extracted in parallel. Equipment for SPE in the 96-well format can be purchased from several different manufacturers. The theory discussed in the SPE sections in this chapter also applies to 96-well SPE.

6.6   Dilute and Shoot

DAS (Figure 6.21) is a simple way of preparing biological samples before injection into an analysis instrument, for example LC-MS. Whereas the traditional sample preparation methods LLE and SPE involve extraction and isolation of analytes from a complex sample, and PPT removes possible interferences from the sample, DAS is only a dilution of the sample with a proper solvent. In other words, compounds are neither isolated nor removed from the sample. This means that interfering compounds still are present in the sample at the injection; however, the concentration of the compounds is greatly reduced.

The idea about DAS is taking advantage of the improved sensitivity and selectivity of modern MS instruments. The sensitivity compensates for the dilution of the analytes, and the concentration of any possible interference causing matrix effects is reduced.

The trend of using DAS-LC-MS, particularly in doping control and analytical toxicology, is often driven by economic benefits. Easy sample preparation and omission of time-consuming extractions are factors that favor the use of DAS. The easiness of the sample preparation is also advantageous in quality assurance as the sample manipulation is reduced, thereby minimizing the total uncertainty in the procedure.

DAS is more frequently used in urine sample preparation compared to plasma and whole blood sample preparation. As described in Chapter 3, urine contains lots of organic and inorganic ions that can cause problems in the LC-MS if the urine sample is injected directly. For example, sediments from crystallization of inorganic salts like calcium phosphate and calcium oxalate can clog the chromatographic system. A simple dilution of the urine sample and hence lowering of the concentration of the inorganic salts can prevent such problems. However, if there still are crystals present in the sample subsequent to dilution, the particulates can easily be removed by centrifugation or filtration. Another possible challenge could

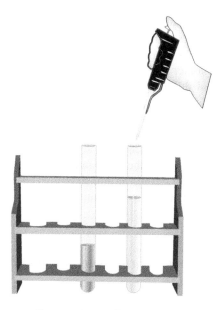

Figure 6.21   *Dilute and shoot*

be proteinuria, a condition where an excess of serum proteins is presented in the urine. Proteins in the urine, which normally should be almost protein free, are likely to cause matrix effects in the MS. Again, a dilution of the urine sample will reduce the probability for such challenges.

There are some examples on DAS with human plasma as the matrix. However, due to the relatively large protein content in such samples, PPT on the column or matrix effects in the MS are likely to occur. Other endogenous components like phospholipids also still remain in the sample after DAS preparation and would probably cause ion suppression or ion enhancement in the MS. Therefore, plasma is still a challenge in combination with DAS. It is suggested to use a monolithic column or an internal surface reversed-phase column as a part of the LC-MS system in order to remove endogenous interferences in an effective way prior to the MS. However, this is a complicated task that should be the subject of more research before it is implemented in the routine laboratory.

DAS would probably minimize the matrix effects in MS but also leads to reduced analyte detectability. Hence, DAS-LC-MS methods still focus on substances for which the required urinary detection levels are high and that show efficient ionization. However, other substance classes, including anabolic androgenic steroids or peptide-based drugs, will probably also be analyzed by DAS-LC-MS with increasing instrument sensitivity.

The advantages of the DAS approach are primarily its simplicity, its cost- and time-effectiveness, and its general compatibility with reversed-phase chromatographic conditions. Because of the simplicity, DAS is easily automated, which again increases the throughput in the laboratory. In addition, the uncertainty of the method is reduced because of the reduced sample manipulation. Then, quantification of substances in urine samples is highly achievable with the DAS method in combination with, for example, LC-MS.

Because all compounds originally found in the sample still are present in the injection aliquot, DAS is a nonselective sample preparation method. Although some of the interferences are diluted below their limit of detection, the lack of clean-up capacity could potentially be a challenge in preparation of complicated or dirty matrices. Insoluble and involatile materials can foul different parts of the analysis instruments, requiring high degrees of cleaning and maintenance and hence excessive instrument downtime.

The simple character of DAS performance only requires dilution of the sample, often urine sample, with a proper solvent that makes the diluted sample compatible with an analysis instrument. How many times the samples should be diluted depends on a compromise between the instrument sensitivity and the compound concentration. The dilution proportion varies in the range from 1:1 to 1:200. The diluent is often pure water, pH-adjusted aqueous buffers, or aqueous solutions with a small content of organic modifier. One example of a DAS method is shown in Box 6.8.

Box 6.8 Dilute and Shoot of a Urine Sample Containing Amphetamine

Background

Amphetamine is a potent central nervous system stimulant and is used in treatment of attention deficit hyperactivity disorder (ADHD) and narcolepsy. Amphetamine is also used by some athletes for its physiological and performance-enhancing effects and is therefore prohibited by the World Anti-Doping Agency (WADA). Because amphetamine is a powerful stimulating drug that produces many effects and can be physically and mentally addictive, it is also a relevant drug in forensic science. The structure of amphetamine is illustrated here:

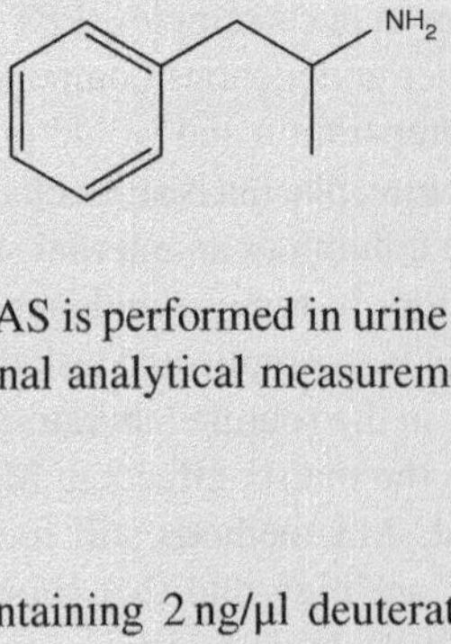

In the procedure presented, DAS is performed in urine samples from patients treated with amphetamine, and the final analytical measurement (not discussed here) is by LC-MS.

Procedure

50 µl of internal standard containing 2 ng/µl deuterated amphetamine is added to 200 µl of the urine sample, followed by 1800 µl of pure water. The sample is vortex mixed and centrifuged for 10 minutes, and 3 µl of the aliquot is injected into LC-MS/MS.

Discussion

The simplicity of the procedure allows high throughput and improved turnaround times. By the present procedure, it is estimated that a batch of 60 samples is prepared in 2 hours, whereas a standard SPE/evaporation/derivatization procedure takes around 8 hours for the same batch size. The decreased sample preparation time is, of course, cost reducing. In addition, the need for expensive equipment, for example SPE cartridges, is almost eliminated. Because of the lack of sample clean-up capability, the DAS approach can potentially lead to dirty instruments and higher operational costs due to increased need for maintenance. However, this can be reduced by extensive use of guard columns.

6.7   What Are the Alternative Strategies?

6.7.1   Dialysis

In *dialysis*, molecules are separated based on molecular size and/or weight. In a standard dialysis experiment, two compartments are separated by a *semipermeable membrane*. Transport across the membrane is based on size exclusion: sufficiently small molecules can permeate the membrane, whereas larger molecules cannot. The principle of dialysis is, as such, based on filtration and not extraction. *Molecular weight cutoff* (MWCO) is used to describe the exclusion limit, and in dialysis the MWCO typically is between 0.1 and 100 kDa. The MWCO is determined as the molecular size/weight where 90% of a molecule of size equal to MWCO is retained. In bioanalysis, dialysis most commonly is used in analysis of plasma or serum samples to separate analytes (small drug molecules) from plasma or serum proteins and other macromolecules (Figure 6.22). A dialysis membrane with an MWCO of 15 kDa is then often applied as proteins that are likely to interfere with the subsequent chromatographic analysis will be retained on top of the dialysis membrane, whereas the smaller drug molecules will pass through the membrane. The dialysis membrane typically consists of regenerated cellulose, cellulose acetate, or polycarbonate, all having a sharp MWCO and relatively low tendency for protein binding.

Dialysis can also be used to retain large molecules of interest on top of the membrane, and as a rule of thumb an MWCO half the size of the molecules to be retained should then be used. Correspondingly, to ensure complete passage of a molecule, an MWCO twice the molecular size should be chosen. The rationale for this is that the permeability of an analyte also is dependent on, for example, molecular shape and polarity.

Dialysis can be performed in the static or dynamic mode. *Equilibrium dialysis* (described in this chapter) is a typical example of dialysis performed in the static mode, keeping both sample and dialysate stagnant. In the dynamic mode, flow is used to increase the dialysis rate. *Dynamic dialysis* can be performed by moving the sample and/or the dialysate. This mode is commonly used in microdialysis (also described here). Dynamic dialysis is convenient if the sample volume is large or if exhaustive dialysis is the goal. Both of the mentioned forms of dialysis result in a sample dilution.

In *microdialysis*, small-diameter hollow dialysis fibers are used instead of flat membranes. In bioanalysis, microdialysis is used to study in vivo drug action mechanisms and the biochemistry of animal behavior. The principle is, for example, applied to sample extracellular fluid (ECF) from the brain or to evaluate the penetration of drug substances through

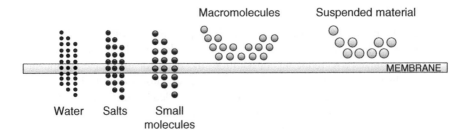

Macromolecules   Suspended material

Water   Salts   Small
molecules

MEMBRANE

Figure 6.22   *Principle of dialysis*

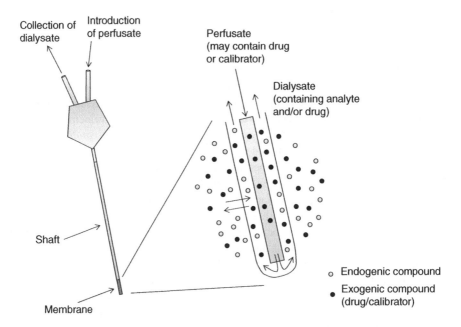

Collection of   Introduction
dialysate   of perfusate

Perfusate
(may contain drug
or calibrator)

Dialysate
(containing analyte
and/or drug)

Shaft

Membrane

o Endogenic compound

• Exogenic compound
(drug/calibrator)

Figure 6.23   *Schematic drawing of a microdialysis probe*

the skin by placing the microdialysis probe under the skin. The technique requires introduction of an ultrathin, semipermeable tube (probe) in the tissue (Figure 6.23). A precision pump is connected to the probe, pumping a tissue-compatible fluid through the probe at low flow rates (1–5 µl/min). Sometimes, this fluid contains calibrators (to determine in vivo recovery), or it may contain drugs to be tested. Small molecules in the ECF surrounding the probe will diffuse passively across the membrane and enter the probe. This includes the non-protein-bound fraction of drugs present. The dialysate is collected at regular intervals and can be analyzed either on- or off-line.

Equilibrium dialysis is a specific dialysis setup used to determine the drug–protein binding in plasma. In this setup, dialysis is performed in the static mode until equilibrium is obtained across the membrane. Plasma is then placed in one compartment, and a buffer-mimicking plasma, usually isotonic phosphate-buffered saline (PBS, pH 7.4), is placed in the other compartment. As the drug–protein binding is temperature dependent, a thermostated dialysis cell is used. Dialysis is then performed until equilibrium (drug concentration in buffer compartment = free drug concentration in sample compartment). As a result, at equilibrium the drug concentration in the plasma is lower (because the free (unbound) drug is distributed in both compartments), and therefore, in order to determine the degree of drug–protein binding or the unbound fraction, the concentration of drug in the buffer compartment as well as the total drug concentration in the plasma compartment must be determined. The unbound fraction can then be calculated by dividing the concentration in the buffer compartment by the total drug concentration. The free drug concentration can then further be calculated by multiplying the unbound fraction with the total drug concentration. Determination of free drug concentration instead of total drug concentration can be of interest for drugs that are highly bound to plasma proteins and that exhibit a variable free concentration (e.g., valproic acid and phenytoin).

Equilibrium dialysis has been regarded as the reference method for determination of drug–protein binding. The equilibrium time must be determined for each compound and in each setup. Equilibrium dialysis is inexpensive and easy to perform, but relatively long equilibrium time is required (4–24 hours). The technique is known to be laborious, but this has been improved by introduction of multiwell plates. Nonspecific adsorption to the dialysis membrane is described, but in equilibrium dialysis this is believed not to affect the unbound fraction.

6.7.2   Ultrafiltration

Ultrafiltration is another way of removing particles from a solution using a semipermeable membrane. The process is, in contrast to dialysis, driven by pressure or centrifugal force. Similar as for dialysis, however, the membrane allows for passages of molecules smaller than the exclusion limit, and in ultrafiltration the MWCO normally is between 1 and 1000 kDa.

Ultrafiltration may, as dialysis, be performed using both flat membranes and hollow fibers. There are both single-unit devices available (100 µl to several milliliters) and multiwell plates.

In bioanalysis, ultrafiltration has similar usage as dialysis. It is used as a method to remove proteins or other large macromolecules without precipitation. Virtually all plasma proteins can be removed, and the sample is (as opposed to PPT) not diluted during sample processing. Ultrafiltration is, together with equilibrium dialysis, the most frequently used technique to determine the degree of drug–protein binding in plasma in early drug development, as ultrafiltration also allows for studying the protein binding without disturbing the equilibrium. The main advantage of ultrafiltration compared to equilibrium dialysis is the speed, especially when using hollow-fiber membranes. It is also easy to adjust pH and temperature to physiological conditions prior to the experiment and to maintain pH and temperature during sample processing. When ultrafiltration is used to determine the degree of drug–protein binding, normally up to 50% of the sample volume is filtrated. Ultrafiltration may also (as dialysis) be used as a sample preparation technique in therapeutic drug monitoring of free drug concentrations.

A potential disadvantage of ultrafiltration is binding of the analyte to the membrane, which results in lower concentration of the analyte in the filtrate. Another disadvantage affecting the volume flux is accumulation of a protein layer on the membrane surface (also called the concentration polarization layer). The formation of this layer is pressure dependent, and the effect on the volume flux increases with increasing pressure applied.

In recent years, another application of ultrafiltration has emerged in analysis of biopharmaceuticals. Protein drugs can be concentrated on top of the filter using an MWCO lower than the molecular weight of the protein to be concentrated. This is also denoted *desalting*.

6.7.3   Affinity Sorbent Extraction

Affinity-based sorbent extraction is extraction based on selective interaction between sorbent material and analyte. There are different kinds of selective sorbents based on the affinity principle available, and in the current chapter three different principles will be described: *antibodies*, *molecularly imprinted polymers* (MIPs), and *aptamers*. All three principles are based on *molecular recognitions*, and several different interactions are important to achieve this, such as hydrophilic interactions (hydrogen bonding and dipole interaction), hydrophobic interactions (van der Waals forces), as well as stacking interactions.

The most widespread affinity principle is the use of sorbent material involving *antigen–antibody interaction*. Using this principle, antibodies are normally covalently bound to an appropriate sorbent forming a so-called *immunosorbent*. This immunosorbent is then packed into a SPE cartridge or pre-column. The sorbent may also be used directly in solution if it is a *coated magnetic particle*, using a magnet to separate the sorbent material from the solution.

Antibodies are synthesized by initiating an immune response in an animal toward the analyte (antigen). The selectivity is dependent on several parameters where antibody type (monoclonal, polyclonal, and fragments) is most important. Antibodies were first commercially developed for large molecules. Antibodies toward small molecules are more recent, and in order to initiate an immune response toward small molecules and analytes, the small molecule has to be bound to a carrier molecule (normally, a protein).

After production, the antibodies are purified and immobilized to a solid support (i.e., silica, agarose, cellulose, or synthetic polymer). The antibodies are normally covalently bound to an epoxide or aldehyde group on the support via the free amino group on the antibody (Figure 6.24). After immobilization, characterization of the immunosorbent is performed with respect to several parameters, such as bonding density, binding capacity, binding specificity and cross-reactivity, breakthrough volume, and recovery.

Immunosorbents are used in the same way as conventional sorbents for SPE, but storage and conditioning are performed in or by using PBS. Antibodies are highly selective and

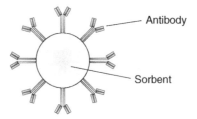

Figure 6.24 *Schematic drawing of an immunosorbent*

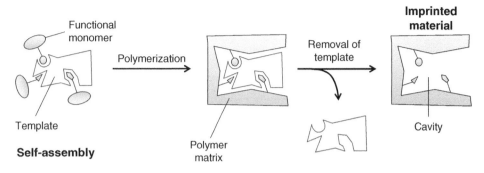

Figure 6.25   *Schematic overview of molecularly imprinted polymer production*

have high affinity toward the analyte used to initiate the immune response. Due to this, the immunosorbent can be used to extract and isolate the analyte from biological matrices in a single step, simultaneously avoiding co-extraction of other matrix components.

MIPs are synthetic polymers with specific cavities for the target analyte. They are sometimes called *synthetic antibodies*, and their main advantages compared to conventional antibodies are with respect to their preparation, which is easy, inexpensive, and rapid. In addition, MIPs provide high thermal and chemical stability.

Several preparation methods for MIPs are described. The most widely used method is the so-called bulk polymerization where the general steps are (i) production of a monolith (with specific cavities for the analyte; see Figure 6.25), (ii) grinding of the monolith to produce particles, and (iii) fractionation of the particles by sieving to isolate the desired particle size range. The procedure is simple, but a disadvantage is that the production results in particles of irregular shape and size. This may make it hard to pack the SPE column efficiently. The whole process is also, despite the simple procedure, relatively time consuming and wasteful because only 30–40% of the produced material is recovered as a usable material.

MIP selectivity is highest in the synthesis medium, and this is usually an aprotic or nonpolar solvent. This has made applications to aqueous biological matrices less straightforward. Lately, focus has therefore been on development of synthesis in aqueous or polar media, making the selectivity better in aqueous sample media. Other approaches to increase selectivity have also been tested. for instance combinations of different monomers.

Aptamers are short (≤ 110 bp) synthetic *single-stranded oligonucleotides* (DNA or RNA). They are prepared by combining a random pool of oligonucleotide sequences from a library with the analyte in question, followed by selection of the sequences that bind. This is done by several iterative cycles of complex separation. The aptamer showing highest binding specificity and affinity is then subsequently isolated and amplified by polymerase chain reaction. Target analytes range from small molecules (100 Da) to large biomolecules (proteins), and aptamers also can be produced toward whole cells and viruses. The whole process of aptamer selection and production is called SELEX (systematic evolution of ligands by exponential enrichment). It is an automatable process, and reasonable amounts of highly specific aptamers for the desired analyte can be produced.

Aptamers for small molecules were introduced recently, and, similar as for production of antibodies, small-molecule analytes have to be linked to a carrier molecule (protein) prior

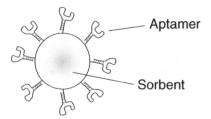

Figure 6.26   *Schematic drawing of an oligosorbent*

to aptamer selection. Only a few aptamers toward small molecules have been produced, characterized, and used for SPE. The combination of an aptamer with a sorbent is called oligosorbents (see Figure 6.26). The oligosorbent is produced by linkage of the aptamer via a spacer to a solid support (e.g., silica, sepharose, or agarose) or to magnetic particles. The produced material is characterized with respect to retention, selectivity, and capacity prior to use.

Aptamers have some advantages over immunosorbents and MIPs. The production is relatively fast and inexpensive compared to that of immunosorbents, and there is no need of laboratory animals. In addition, only small amounts of template are necessary, which is an advantage compared to MIPs. The use of aptamers in bioanalytical applications is still limited.

6.7.4   Solid-Phase Microextraction

Solid-phase microextraction (SPME) was introduced in the early 1990s as an almost solvent-free microextraction technique based on partitioning and extraction of an analyte between two immiscible phases, a solid extraction phase and an aqueous sample. The SPME process basically consists of two steps: (i) partitioning of the analyte between the matrix and the extraction phase, depending on the partition ratio of the analyte; and (ii) desorption of the analyte from the extraction phase and into the analytical instrument. Traditionally, SPME mostly has been used in combination with GC, where thermally desorption of volatile analytes can occur. However, there is an increased trend for combining SPME with liquid chromatography in order to analyze weakly volatile or thermally labile compounds not amenable to GC. If SPME is used in combination with LC, the analytes are desorbed either in an online SPME–LC interface or offline into a desorption solvent, which is subsequently injected into the LC instrument by an autosampler.

Conventional SPME devices consist of a *fused-silica fiber* coated on the outside with an appropriate stationary phase, so-called fiber-SPME. The device basically consists of an assembly holder and a fiber assembly that protects the built-in extraction fiber, as shown in Figure 6.27. The stationary phase can be a liquid (polymer), a solid (sorbent), or a combination of them.

Opposed to conventional sample preparation techniques like LLE and SPE, SPME is an equilibrium sample preparation method, where only a small percentage of the analyte is removed from the sample and extracted into the small extraction phase. Some of the features that characterize SPME include efficient sample clean-up; it can be applied on small sample volumes (<100 μl); almost eliminated use of hazardous organic solvent; and

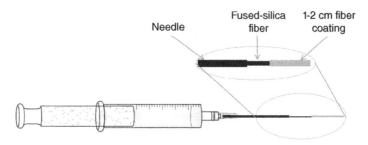

Figure 6.27   *Illustration of a fiber-solid-phase microextraction device*

simplification of the general sample preparation procedure by minimizing the number of steps required.

SPME can be operated in different extraction modes, depending on the nature of the analyte and the matrix. In *head space solid-phase microextraction* (HS-SPME), which is the mostly used extraction mode in SPME, the fiber is positioned in the head space above the matrix. The matrix can be either an aqueous or a solid sample. Volatile analytes in equilibrium between the matrix and the head space will be extracted by the extraction phase. One of the advantages by using HS-SPME is that nonvolatile, high-molecular-weight matrix interferences never are in direct contact with the extraction phase, hence increased selectivity can be obtained. In addition, possible fouling problems from the matrix are minimized. In the *direct extraction solid-phase microextraction* mode (DI-SPME), the SPME fiber is immersed into the aqueous sample and allows the analytes of interest to partition between the sample and the extraction phase on the fiber. A magnetic stirrer is often situated in the sample in order to increase the convection of the sample and hence increase the equilibrium rate. However, the direct contact between the sorbent and the matrix can lead to fouling of the extraction phase and hence reduced sensitivity, reproducibility, accuracy, and extract integrity from complicated matrices like biological samples. A third mode that should be mentioned is the *membrane-protected SPME* mode. The main purpose of this rarely used mode is to protect the fiber against damage when the fiber is immersed in a dirty sample.

There are many different extraction phases commercially available. Their mutually different physicochemical characteristics help to increase the application range of SPME. Fiber coatings are available in increasing thicknesses from 7 to 150 μm, which increases the extracted amount of the analytes—and, hence, analyte detectability—but also increases equilibrium times. The coatings include single-phase adsorbents like the nonpolar polydimethylsiloxane (PDMS), more polar polymers like polyacrylate (PA), and Carbowax (CW) and mixed-phase sorbents in which nonpolar and polar polymers are mixed together.

Increased thickness of the sorbents means longer extraction times. This can be compensated for by increasing the surface area by using a thin-film geometry. Such a thin-film SPME provides more extracted analytes without sacrificing time. However, the extraction principles still remain the same as discussed for fiber-SPME. The thin-film configuration allows SPME to be performed in a multiwell format. Another configuration of SPME is the in-tube SPME, where the analytes of interest are adsorbed to the inner wall of a capillary tubing. This configuration was mainly intended for automation and high-throughput performance of SPME.

The main advantages of SPME are its simplicity, solvent elimination, high sensitivity, small sample volumes, relative low costs, and simple automation. These features are of course of big importance in bioanalytical sample preparation. Although the focus for the first years of SPME was on extraction from relatively simple matrices, for example environmental samples, interest in the combination of SPME and complex biological matrices has grown significantly. There are lots of examples of SPME methods developed to assay compounds in biological samples, including urine, serum, plasma, whole blood, saliva, and hair.

Analytes in biological samples are extracted either by the HS-SPME technique for volatile analytes or by the DI-SPME followed by derivatization for less volatile analytes. In both cases, the extraction is often followed by analysis on GC-MS. In such procedures, low limits of detection (LODs) and excellent quantitation are obtained. However, for more polar drugs and metabolites in plasma, the application of SPME may be limited to compounds with relatively high therapeutic concentrations due to the relatively low partition ratio between polar analytes and the SPME coatings and hence relatively high LODs.

6.7.5   Membrane-Based Extractions

Other sample preparation techniques that are worth mentioning are the membrane-based extraction techniques. They share common features in which a thin membrane separates two phases, and enables transport of analytes of interest from one phase, across the membrane, and further into the receiver phase on the opposite site of the membrane. The membranes can be classified as either porous or nonporous. The selectivity of the porous membranes is mainly based on pore size and pore size exclusion, and has similarities with ultrafiltration (Section 6.7.2). The technique will therefore not be mentioned anymore in this section. The nonporous membranes can be either a porous membrane impregnated with an organic solvent to create a so-called *supported liquid membrane* (SLM), or entirely a solid, for example silicone rubber. The solid membranes are not frequently used in bioanalytical sample preparation and will not be mentioned anymore in this section. The transport across liquid membranes can be facilitated by passive diffusion driven by a pH gradient, by an electrical potential difference across the membrane, or by an applied pressure across the membrane. The two latter techniques are not frequently used in a bioanalytical laboratory and will therefore not be discussed anymore.

The membrane extraction method most frequently used in bioanalysis is based on SLMs. SLMs have the advantages of a high degree of selectivity and clean-up efficiency, enrichment capability, and minimized use of organic solvents. The extraction principle is similar to the traditional theory of LLE, where analytes are partitioned between an aqueous solution and an organic solvent based on the partition ratio of the analyte (Section 6.4). In SLM extractions, the organic solvent is incorporated into a thin, porous membrane. Features of the organic solvent include immiscibility with water and a high boiling point in order to avoid evaporation. The SLM creates a physical barrier between two solutions: a sample solution and an acceptor solution. If the acceptor solution is an organic solvent, it is called a *two-phase system* and follows common LLE principles. If the acceptor solution is aqueous, a three-phase system is present, and the principle is similar to LLE with back

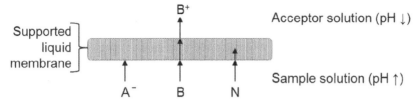

Figure 6.28 *Schematic illustration of a supported liquid membrane extraction process exemplified by a basic analyte*

extraction. A schematic illustration of a *three-phase system* is shown in Figure 6.28. In the example, the pH in the sample solution is high, resulting in charged acids (A), uncharged bases (B), and neutrals (N). Only the basic and neutral compounds will be extracted into the SLM. The pH of the acceptor solution on the opposite side of SLM is low; hence, the partitioning of the basic compound from the SLM and into the acceptor solution is favored.

Due to the extraction mechanism described above, an SLM extraction is a selective extraction method with a high degree of sample cleanup capability. This makes it a valuable sample preparation tool in bioanalysis, where the matrices often are complicated and contain lots of components that can interfere with the analysis result. However, SLM extraction equipment is not yet commercially available, and this sample preparation method is therefore still not used in clinical routine laboratories.

The SLM extractions in bioanalysis are mainly performed in a hollow-fiber configuration (*hollow-fiber liquid-phase microextraction,* HF-LPME) or a flat membrane-based multi-well configuration (*parallel liquid membrane extraction,* PALME).

7

High-Performance Liquid Chromatography (HPLC) and High-Performance Liquid Chromatography–Mass Spectrometry (LC-MS)

Steen Honoré Hansen[1] and Leon Reubsaet[2]

[1]*School of Pharmaceutical Sciences, University of Copenhagen, Denmark*
[2]*School of Pharmacy, University of Oslo, Norway*

High-performance liquid chromatography (HPLC, or for short LC) is the most commonly used chromatographic technique to determine drugs in pharmaceutical preparations and in biological material. In LC, the mobile phase is a liquid forced through a column packed with a material that retards the analytes introduced into the system. The analytes are injected into the flow of mobile phase just in front of the separation column. The outlet of the column is connected to a detector where the eluted substances are detected. The separation principle can be any of the principles described in Chapter 4. A number of detection principles are available, but for bioanalysis, mass spectrometric detection has become the principle of choice. In this chapter, it is reviewed how the apparatus is constructed and how it works.

7.1 Introduction

The names of LC are many. When HPLC was introduced in the late 1960s, the name *high-pressure liquid chromatography* was used. The pressure is, however, not a desire but more an unavoidable drawback, and the name was changed to HPLC, focusing on the very good separations obtained. It is now becoming common practice just to use the shortened

Bioanalysis of Pharmaceuticals: Sample Preparation, Separation Techniques and Mass Spectrometry,
First Edition. Steen Honoré Hansen and Stig Pedersen-Bjergaard.
© 2015 John Wiley & Sons, Ltd. Published 2015 by John Wiley & Sons, Ltd.

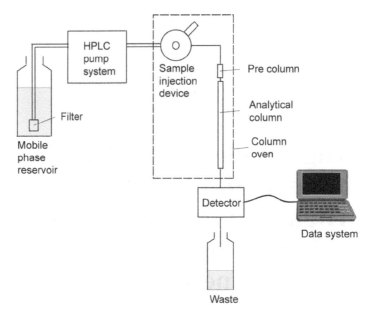

Figure 7.1   *Main structure of a liquid chromatography system*

term LC. Systems capable of running at very high pressures have also been introduced under the name *ultrahigh-performance liquid chromatography* (UHPLC or just UPLC; this is used as a trademark). The main components of an LC system are shown in Figure 7.1.

The separation columns used for bioanalytical purposes are typically made of steel tubes 5–25 cm long and packed with the stationary phase with an internal diameter of about 2–4.6 mm, but also micro- and nano-LC systems are in use. It is in the column that the sample components are separated, and thus the column is the heart of the chromatograph. The other parts of the chromatograph are optimized individually and carefully put together to optimize the separation efficiency of the total system. The three main parts of the LC system—the *solvent delivery*, the *separation column*, and the *detector*—are all vital and indispensable units. The separation occurs when the mobile phase is pumped at a *constant flow rate* through the column bringing the separated analytes to the detector. In a standard bioanalytical LC system, the typical flow rate of the mobile phase through the column is 0.2–0.5 ml/min but can be in the range from 0.01 to 10 ml/min. The small particle size of the column packing materials results in a back pressure of 30–300 bar (3–30 MPa) and in certain cases (UHPLC) up to 1500 bar (about 150 MPa), when the mobile phase is pumped through the column. The pumps used must be able to pump the mobile phase at a constant flow rate against a high pressure. Any particle in the samples injected will be collected on the top of the column and will gradually block the column, with an increase in *back pressure* as the result. This will result in a decrease in flow rate that will compromise the analytical result. Therefore, HPLC pumps are equipped with a regulation mechanism that keeps the flow rate constant, and a gradual blockage will thus result in an increase in back pressure while the flow is kept constant.

The compounds to be separated must be dissolved in a liquid that is miscible with and not stronger eluting than the mobile phase. Typical injection volumes are 1–100 µl. The injection systems are optimized to inject the solution under high pressure directly into the flowing mobile phase just before the column inlet. The detectors provide an electronic response to the analytes. The response is processed by the computer system that displays the results as chromatograms, and for quantitative purposes the area of the peaks is also determined. The whole analysis process can be automated and controlled by the computer system. When an autosampler is used as the injector, the LC system can work 24 hours a day. Analytical chemical data, including quantitative calculations, can be reported continuously by the computer system.

The broad applicability and the high degree of automation of LC are some of the reasons why the technique has gained such a dominant position in analysis of food, pharmaceuticals, and biofluids.

Liquid chromatography–mass spectrometry (LC-MS) has become of special importance in drug research and development because of the high separation efficiency combined with the sensitivity and selectivity of mass spectrometric detection, which together provide very reliable analytical chemical data.

7.2   The Solvent Delivery System

One or more LC pumps may deliver mobile phases from up to four different solvent reservoirs in order to perform either isocratic or gradient elution (see also Chapter 4). The purpose is to deliver the programmed mixture of solvents to the analytical column at a controlled flow rate. The pumps can be constructed in different ways, but the piston pump is the most common. A sketch of the piston pump is shown in Figure 7.2.

The piston pump consists of a small steel cylinder with a volume of approximately 100 µl. A piston is moved back and forth in the cylinder by means of a motor. There are two ball valves (check valves) attached to the cylinder so that the mobile phase can only flow in one direction, into the cylinder from the reservoir or out to the column. When the piston is moved back, the lower ball valve will open while the upper ball valve will close, dragging the mobile phase into the cylinder. When the piston again is moved forward into the

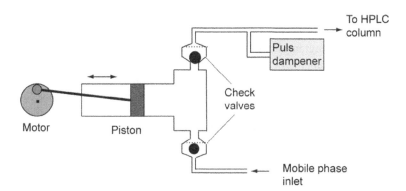

Figure 7.2   *The main parts in a piston pump*

cylinder, the bottom ball valve will close, while the top valve will open. Mobile phase is then forced out of the cylinder through the injector into the column. Because the mobile phase is forced into the column only when the plunger is pushed into the cylinder, the fluid flow will pulsate. The pulsation will introduce extra noise to the detector signal and should be eliminated if possible. A pulse dampener is therefore included in the system to ensure a smooth flow of the mobile phase. Other pump systems ensure a smooth flow by linking together two piston heads into a double piston pump, where one piston head delivers mobile phase to the column while the other is filling up with mobile phase and vice versa.

When the pumping system delivers a mobile phase with a constant composition to the column, it is called *isocratic elution*. However, it is also possible to have a system where liquid is pumped from two or more reservoirs, and then mix them during chromatography. This can be done using a single pump equipped with a low-pressure mixing valve connected to up to four different reservoirs containing liquids. The mixing valve will open for only one pipeline at a time, and in this way the solvent mixture can be controlled. It is also possible to use more than one pump where each pump delivers a controlled amount of each liquid. The mixing is then performed at the high-pressure side, but the two principles are of similar performance.

In *gradient elution*, the eluting strength of the mobile phase is gradually increased during chromatography. This has the advantage that late eluting peaks will be eluted earlier as sharper and higher peaks, thus improving detection limits. The immediate reduction in the observed analysis time is, however, somewhat lost due to the needed time for re-equilibration to the initial mobile phase condition before injection of the next sample. A correct re-equilibration is very important in order to keep constant retention times from sample to sample. If full re-equilibration is not achieved, the retention of the analytes in the following chromatogram will be shorter—especially of the early eluting peaks. Gradient elution in the form of a washing step with strong eluting properties may also be useful in order to "clean" the column for strongly retained solutes. Gradient elution can be compared to temperature programming in GC (gas chromatography).

Biofluids and other biomaterials contain a huge number of compounds, some highly polar and some very lipophilic. Even though sample preparation is performed before chromatography, the sample for injection often still contains compounds with a wide span in polarity and thus very different affinities toward the stationary phase (a large span in distribution constants).

Gradient elution is used to separate samples containing compounds with large differences in retention. When large differences in retention of analytes are current, isocratic elution will result in unnecessarily long analysis time. In addition, it is only the analytes in the middle of the chromatogram that elute satisfactorily: the least retarded substances eluted early as partially overlapping peaks, whereas the most retarded substances eluted as broad peaks with a long retention time. This is illustrated in Figure 7.3.

Using gradient elution, the composition of mobile phase is changed during chromatography, starting with a weaker eluting composition of the mobile phase. In this way, the least retarded substances will obtain sufficient retention and separation. The strength of the mobile phase is then increased during chromatography, and the late elution peaks are now eluted faster and with a better sensitivity. This is also illustrated in Figure 7.3.

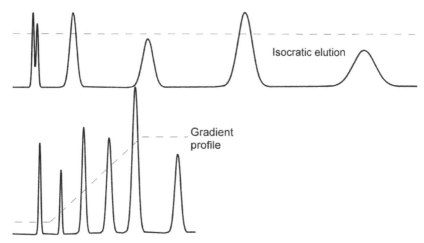

Figure 7.3   *Isocratic and gradient elution of a sample containing analytes with large differences in retention*

7.3   Degassing and Filtering of Mobile Phases

The polar mobile phases used in reversed phase chromatography can dissolve gasses (air), and this may pose a problem especially in gradient elution where the aqueous solvent is mixed with an organic solvent. The solubility of gasses in the mobile phase increases with increasing pressure, but when the pressure again decreases along the column, the air may be released as small air bubbles that can disturb the detector signal. It is therefore necessary to degas the mobile phases. This can be done in several ways: by refluxing, by replacement with a less soluble gas, by ultrasonic treatment, or by vacuum. At present, on-line vacuum degassing is the preferred way. The inlet line from the solvent reservoir is connected to the on-line degasser where the mobile phase is passing through a thin-walled porous polytetrafluoroethylene (PTFE) tubing placed in a vacuum chamber. The vacuum "pulls" the air out of the mobile phase, while the mobile phase is kept within the tubing because of the high hydrophobicity of the tubing. In this way, a constant degassing is obtained.

When using UV detection in the low-UV range below 220 nm for trace analysis, thorough degassing is also needed because oxygen absorbs at these wavelengths, resulting in a higher background signal and loss of sensitivity. Using electrochemical detection in the reduction mode will also require removal of oxygen.

Filtering of mobile phases is not always necessary. Distilled water or deionized water from Milli-Q^{TM} equipment and organic solvents as well as mixtures thereof are used as supplied. But sometimes when buffers have been prepared, some precipitate or mechanical impurities can be observed. Such solvents should be filtered before they are used for HPLC. The inlet tubing from the solvent reservoir to the HPLC is furthermore normally equipped with a filter. This is to avoid any particles from entering the pump, where impurities on the sapphire balls in the check valves can prevent them from correctly closing and thus will result in malfunction of the pump.

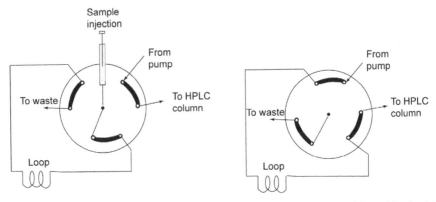

Figure 7.4   *A two-position, six-port injection valve in the load position (a) and in the inject position (b)*

7.4   Injection of Samples

Samples have to be introduced into the system at a high back pressure. Therefore, all injection valves have to be of a high quality to be able to operate for thousands of injections without malfunction. A typical two-position, six-port valve is shown in Figure 7.4. These valves are used for manual injection when using a syringe, but the same principles is used in auto-samplers. In the load position, the mobile phase from the pump passes through the valve directly to the column. In this position, it is possible to inject the sample into a loop using a syringe. When the valve then is switched to the inject position, the mobile phase from the pump will pass through the loop and in this way bring the sample to the column.

When the loop is only partially filled with sample, it is important that the loop is back flushed to the column to prevent unnecessary mixing. The valves will become worn after a certain amount of use, depending on the quality of the samples and the mobile phases used. When the rotary plates with the tiny channels in the valve get scratched, due to particles coming from samples, some sample may leak in between the rotary plates and will show up as a carryover between samples. Replacement plates have then to be installed. It is therefore of vital importance to include injection of blank samples and/or injection of mobile phase with regular intervals when running samples. In some instruments, automatic washes of syringes are installed and can be programmed to carry out wash between each sample. However, such a system can also be contaminated and may therefore have to be cleaned. Using an auto-sampler, the process can be controlled by the HPLC software so that analysis can take place without supervision running for 24 hours or more. Auto-samplers can also be equipped with a refrigerator for cooling of sensitive samples so that they do not decompose before analysis.

7.5   Temperature Control

Temperature control is of vital importance throughout the bioanalytical process because of the possible instability of samples. In this section, only the physical aspects while the sample molecules pass the column are discussed. It is important to control the temperature of the column because the partition ratio is influenced by changes in temperature. The raise in column temperature from ambient to 40 or 50 °C can have several effects on the chromatography. First of all, it is important to verify whether the column can be used at higher temperatures. Older C18 column packing materials were often not recommended to be used above 60 °C, but some of the new materials can be used up to at least 90 °C. The limitation is primarily due to hydrolysis of the silica polymer and the resulting loss of stationary phase. A higher column temperature will also reduce the viscosity of the mobile phase and an improved mass transfer is the result, which often results in an increase in column efficiency. It is, though, important to be aware of temperature differences between the sample and the mobile phase. If a major difference is present, band broadening may take place due to mixing, resulting in loss of efficiency. This problem can at least partly be overcome with preheating of the mobile phase to the same temperature as the column temperature. When analyzing larger molecules like proteins hydrolysis, changes in conformation and aggregation can be a result of an increased temperature.

7.6   Mobile Phases

The following requirements for liquids to be used for mobile phases have to be considered:

- The solvents should preferably not give any response in the detector used.
- The solvents must have a satisfactory degree of purity. A number of solvents are available in so-called HPLC quality and sometimes are specified for selected purposes (e.g., for gradient elution or for MS detection).
- The solvents should have low viscosity to provide as low a back pressure as possible in the HPLC system.
- The solvents should have low toxicity, preferably are inflammable and nonreactive, and must be suitable for disposal after use.

Mobile phases for reversed phase chromatography often contain aqueous buffer solutions. The buffer salts are dissolved in water of HPLC quality. Water of HPLC quality is produced in the laboratory using a water treatment plant that removes common contaminants from tap water, or it can be obtained commercially. The aqueous solution is mixed with an organic modifier to the prescribed solvent strength. If the buffer salts give rise to particles in the solution, it has to be filtered through a filter with a pore size of 0.45 µm (0.2 µm for UHPLC). Before use, the mobile phase has to be degassed to remove dissolved air as described above.

Viscosity of the mobile phase has to be considered because the back pressure in the LC system will increase with increasing viscosity. In Figure 4.18, the viscosity of a number of organic solvents with water is shown as a function of the water percentage. Mixtures of methanol and water become more viscous at 40–50% methanol, and if high back pressure is a problem, acetonitrile can be used as an alternative modifier.

The UV detector is, in general, by far the most widely used HPLC detector. Analytes can be detected when they have a UV response above 190 nm. This requires the mobile phase to be transparent at the given wavelength. Table 7.1 gives an overview of the *UV cut-off* of common solvents. Most often, methanol and acetonitrile are used as the organic

Table 7.1   *UV cut-off for common solvents (1 cm path length)*

Solvent	UV cut-off (nm)
Acetone	330
Acetonitrile	190
Dichloromethane	220
Ethanol	210
Heptane	195
Methanol	205
Tetrahydrofuran	210
Water	190

The UV response increases as the wavelength decreases, and the UV cutoff value is the shortest wavelength that can be used with the solvent.

modifier due to their low UV cut-off, but also tetrahydrofuran (THF) can be used. Acetone is also miscible with water but is not used as often due to its UV cut-off at high wavelength. However, when MS is used as the detection principle, this will be of no concern and acetone has a solvent strength similar to that of acetonitrile. In bioanalysis, MS has become the detection principle of choice, and this put some special requirements on the mobile phases used (see also Section 7.9).

It is in general wise to control the pH in the mobile phases even if only neutral solutes are to be analyzed because also the stationary phase has to be controlled, and the sample may contain ionizable impurities whose retention time is pH dependent. When using UV detection, it is preferable that the mobile phase has a low UV background, and therefore phosphate buffers have been preferred due to the lack of UV absorbance. MS detection is performed in high vacuum, and as a result the components of the mobile phase have to be volatile while any UV-absorbing characteristics are of less importance. This is discussed in more detail in Section 7.9.

7.7   Stationary Phases and Columns

The separation column (for HPLC and UHPLC) is the heart of the chromatographic system. It is in the column that the analytes distribute between the two phases and in this way get separated. It is the composition of the mobile phase and the nature of the stationary phase that primarily control the separation. The separation principle is either straight phase (normal phase) chromatography, where the stationary phase is more polar than the mobile phase, or reversed phase chromatography, where the stationary phase is more nonpolar than the mobile phase (see also Chapter 4).

HPLC has been commercially available since the late 1960s, and it was a major break-through when stable stationary phases to be used in reversed phase chromatography became available around 1970. The most popular setup is silica with chemically bonded station-ary hydrophobic phases used together with aqueous mobile phase made up of water and a miscible organic solvent like methanol or acetonitrile. The mobile phase can be fur-ther modified with a buffer to control pH and also with other additives. Reversed phase

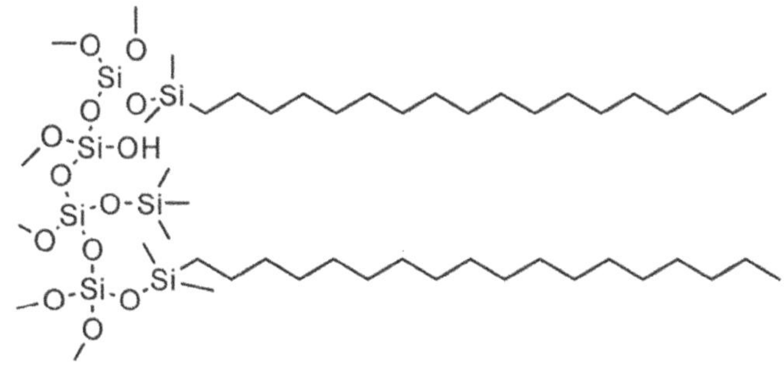

Figure 7.5 *Octadecylsilylsilica (ODS column packing material) for reversed phase chromatography*

chromatography made it possible to directly analyze aqueous samples, and this principle has now become by far the most used HPLC technique of all.

Basically, the chemically bonded phase packing materials consist of silica material chemically modified with more or less hydrophobic organic molecules. The most used material is octadecylsilylsilica (called ODS or C18 material; Figure 7.5), which has C18 hydrocarbon chains on the surface. There seems to be no limit to the type of modification of the silica surface that can be performed. Presently (2014), there are close to 1000 different C18 materials on the market, and they are all more or less different in their chromatographic characteristics because of differences in manufacture of the silica and also in the bonding of the C18 groups. It is therefore recommended not to change the brand of a column packing material when a method first has been established.

A column packing material based on silica is not fully stable because very small amounts will dissolve over time. This is generally not a problem at low pH, but at pH above about 8.5 silica dissolves and will disappear if no precautions are taken. However, from the beginning of this millennium, a number of column packing materials for reversed phase chromatography that can be operated at high pH have become available.

A tremendous development in technology has taken place in the manufacturing of column packing materials. This again has necessitated a technological development in HPLC equipment. Starting out in about 1970 with *solid core particles* of about 40 μm in diameter having a thin layer of porous silica on the surface, smaller, *fully porous particle* sizes of 10 and 5 μm soon became available (Figure 7.6).

In the 1990s 3 μm became available, and now column packing materials with particle sizes of 1.5–2 μm are state of the art. It is remarkable that a number of the small particles now again are manufactured as spherical particles with a solid core and with a thin porous layer on the surface. The decrease in particle diameter will of course lead to higher back pressures in the system. This can be overcome in two ways: a shorter column can be used because the smaller particles give more efficiency (a higher number of theoretical plates), or by operating at higher pressures. The need for pumps able to deliver mobile phases against the much higher back pressures resulted in new pump technology and also improved injection devises and detection systems. To fully exploit the use of small particles (≤2 μm), this special equipment (termed UHPLC) is needed.

Columns packed with 5 and 3 μm particles can be run in a standard HPLC system with a pressure limit of 400 bars. But 1.5 μm particles require equipment that is able to run

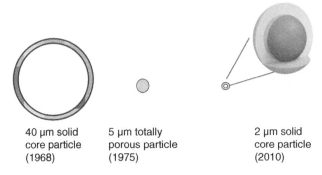

40 μm solid   5 μm totally   2 μm solid
core particle   porous particle   core particle
(1968)   (1975)   (2010)

Figure 7.6   *The development in column packing materials*

at a much higher back pressure, and therefore UHPLC instruments were developed. The chromatographic principles are the same as previously, but new demand on the pumps, auto-samplers, and detectors had to be met. The pumps and auto-samplers have to be able to withstand high pressures up to 1000–1500 bars. Using the column packing materials with 1.5–2 μm particles will provide a very high number of theoretical plates, and therefore the column length can be reduced and analysis time and amount of mobile phase can be considerably reduced (see Tables 7.2 and 7.3). The high efficiency gives very narrow chromatographic peaks, which means new demand on the detector cell volume to avoid that peaks separated on the column are mixed up in the detector cell. To fully explore the advantages (high efficiency and high speed of analysis) of UHPLC, UHPLC equipment is needed. On the other hand, it is possible to use columns with 2–2.5 μm particles in standard chromatographic systems if the column length is not reduced too much (not below 50 mm). To do this, it is important to optimize (reduce) the *extra column volume*, which is in the connecting tubing from the injection device to the column and from the column to the detector, including the detector cell volume.

Table 7.2   *Effect of column length and particle size on column efficiency (N)*

Column Length (mm)	Resolving Power N(5 μm)	Resolving Power N(3.5 μm)	Resolving Power N(1.8 μm)	Resolving Power N(1.3 μm)
150	12,500	21,000	-	-
100	8,500	14,000	-	-
75	6,000	10,500	-	-
50	4,200	7,000	12,000	-
30	-	4,200	6,500	12,000
15	-	2,100	2,500	-

Typical numbers of N for columns packed with column packing materials with 5, 3.5, 1.8, or 1.3 μm particles.

Table 7.3 *Some typical high-performance liquid chromatography columns and the corresponding eluent consumption*

Column dimension (mm)	Particle size (μm)	Flow rate (ml/min)	Mobile phase consumption (relative %)	Typical injection volume(μl)	Volume of liquid in column (μl) permeability = 0.7
250 × 4.6	5	1.0	100	20	2906
150 × 4.6	3	1.0	60	20	1744
150 × 2.0	3	0.2	8	5	231
100 × 2.0	2.5	0.2	5	5	154
50 × 4.6	1.8	1.0	33	20	581
100 × 4.6	1.8	1.0	60	20	1163

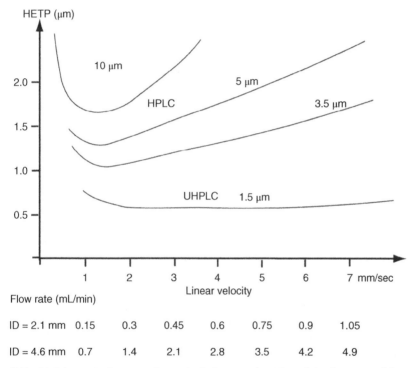

Figure 7.7 *Height equivalent to a theoretical plate as a function of the flow rate of the mobile phase for 1.5, 3.5, 5, and 10 μm particles*

Decreasing particle size will provide a larger number of theoretical plates. This is illustrated in Figure 7.7, where the height equivalent to a theoretical plate (HETP) is plotted versus particle sizes of 1.5, 3.5, 5, and 10 μm.

The optimum efficiency is at the smaller particle sizes shifted toward higher linear flow rates. This indicated that higher flow rates can be used without losing efficiency. However, the small particle size also provides a very high back pressure, and thus shorter column lengths have to be used.

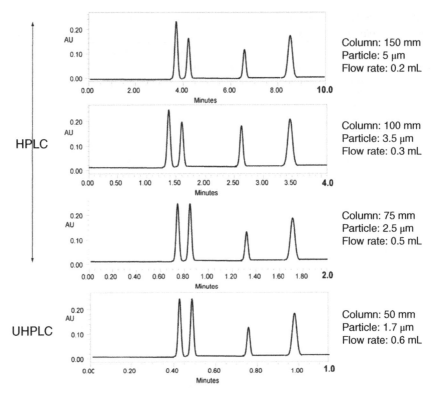

Figure 7.8 *Effect of the reduction in particle size and column length and the analysis time and chromatographic separation*

The practical implication of this is that short columns with particles of about 2 μm can give efficient separations in a very short time (Figure 7.8). Depending on the back pressure, it can be necessary to use UHPLC equipment.

The typical analytical HPLC column has until the mid-1990s been a 15–25 cm long steel tube packed with 5 μm particles. The inner diameter of the tube has been 4.6 mm. With the introduction of MS as a routine LC detection technique and the improvement of column packing for HPLC, column dimensions of 10–15 cm with 2 mm internal diameter have become the common standard. For fast analysis, smaller particles between 1.5 and 2 μm are used in columns of 3–5 cm in length.

The internal diameter of the column has an influence on the size of the detector signal. Reducing the internal diameter from 4.6 to 2 mm, the volume of the column is reduced with a factor of about 5, which will reduce the dilution of an identical injection volume with the same factor. Also the particle diameter will influence the peak width, and this is illustrated in Figure 7.9.

Using shorter columns with smaller internal diameters results in major savings in consumption of mobile phase. In Table 7.3, the reduction in mobile phase consumption is calculated for columns of equal length with different internal diameters for a constant flow rate of mobile phase through the columns.

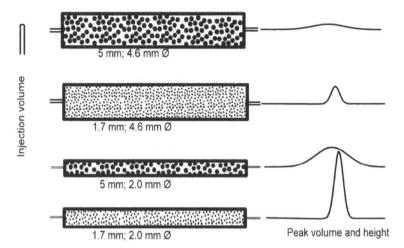

Figure 7.9 *Effect of particle diameter (μm) and internal column diameter (Ø) on peak signal with identical injection volume*

A further reduction of consumption in mobile phase is proportional to the simultaneous reduction in column length. A reduction in column length will also result in a similar reduction in analysis time.

The column packing material is held in place in the column by a metal filter at each end. The metal filter is either a porous frit or a net so that the mobile phase can pass through. The pore size of the filter has to be smaller than the diameter of the packing material in order to prevent particles from leaking from the column.

A narrow particle size distribution of the column packing material is important for a good and uniform packing of the column. This, combined with low dead volume end fittings of the column and short tubing with low internal diameter from the injection device to the column and from the column to the detector, will secure high quality of the separations. If non-optimal connections of tubings and fittings are made when installing the column in the LC system, the separation can be less efficient due to extra column band broadening, even if the column itself is in perfect condition. Remember: when measuring the efficiency of the "column," it is the total system that is being tested.

7.8   Detectors

7.8.1   Introduction

The physicochemical principles behind the many types of detectors available for detection in LC are very different. The UV and fluorescence detectors are based on interaction of the analytes with light, whereas the electrochemical detector is based on electrochemistry (redox reactions). The very popular MS detector can be considered as a kind of a balance, where ions, after separation, are weighted. The radiochemical detection is in the end the measurement of emitted light resulting from the activation of a scintillator by the radioactive β-emission from the radioactive tracer.

Table 7.4   *Some commercial available liquid chromatography detectors commonly used in bioanalysis and their typical performance*

Detector	Lower limit of detection (ng)	Gradient elution?
Ultraviolet (UV)	0.1–1	Yes
Fluorescence	0.001–0.01	Yes
Electrochemical (EC)	0.01–1	No
Mass spectrometry (MS)	0.001–0.01	Yes
Radiochemical	Depends on degree and type of radiolabeling	Yes

The LC detector should give a response for the analyte that is converted into an electrical signal. The response should be proportional to the concentration of either substance in the mobile phase or with the mass of the substance in the mobile phase, so quantitative analysis can be carried out based on measurement of peak areas or peak heights. The detectors can be divided into two types: general detectors that measure any change in the mobile phase, and specific detectors that respond only to substances with specific properties. The most important detectors used in bioanalysis are shown in Table 7.4 and are discussed in the following.

In most HPLC equipment, a UV detector is included as a standard detector. However, in many bioanalytical applications this detector is not sufficiently sensitive. Its selectivity is also limited because many compounds exhibit UV absorption at similar wavelengths. Fluorescence detectors and electrochemical detectors have for selected analytes a much lower limit of detection compared to the UV detector. The mass spectrometer provides additional information on the molecular structure, and as its sensitivity also is high for most analytes it has become the detector of choice for bioanalytical chemistry. Mass spectrometers are treated in depth in Section 7.9.

7.8.2   Spectroscopy

Measurements in bioanalysis are often based on spectroscopy techniques. Measurements can be performed directly on samples, but in many cases separation techniques are involved and here spectroscopy is used for detection of the analytes after the separation (e.g., in HPLC).

The basic principle of spectroscopy is the interaction of light with matter (e.g., analytes). Light can be considered as waves of energy, and a basic definition says that:

$$c = \lambda \cdot \upsilon \tag{7.1}$$

where c is the speed of light, λ is the wavelength of the radiation, and υ is the frequency in hertz. Light can also be considered as a stream of particles (photons) having a certain energy:

$$E = h \cdot \upsilon \tag{7.2}$$

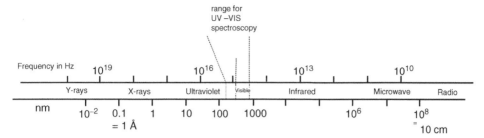

Figure 7.10   *The electromagnetic spectrum, showing the connection between frequency and wavelength*

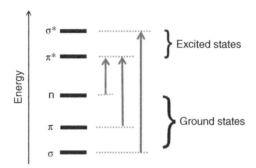

Figure 7.11   *Diagram illustrating light excitation of electron*

where *h* is a constant (Planck's constant), and *υ* again is the frequency. Thus, as the speed of light, *c*, is constant, a shorter wavelength implies a higher frequency and thus a higher energy (Figure 7.10).

In this chapter, we will restrict ourselves to discussing radiation in the ultraviolet and visible region because this is the range used for bioanalysis measurements. In order to use spectroscopy for measurement, the light has to interact with the analytes; or, in other words, the analytes have to absorb light energy. This energy absorption takes place as an exciting of electrons from a ground state to a higher energy state (Figure 7.11).

To excite an electron to a higher energy state, a certain amount of energy is required corresponding to a certain wavelength. Electrons in a sigma (*σ*) bond require more energy and thus a shorter wavelength of the light to be excited. *π* electrons, typically present in double bonds, are more easy to excite, whereas nonbonding electrons (n) usually require even less energy to be excited.

The exciting of *σ* electrons present in single bonds normally does not give rise to absorbance above 200 nm because more energetic light is required. Double bonds present in a molecule will result in UV absorbance above 200 nm, and if the double bond is further conjugated (double bond–single bond–double bond), less energy is required for the excitation. A molecule with many conjugated double bonds will be colored.

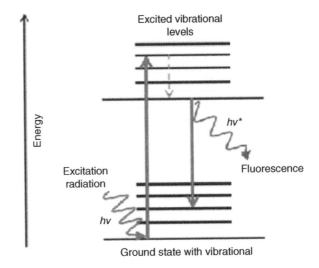

Energy

Excited vibrational
levels

hv*

Excitation
radiation

Fluorescence

hv

Ground state with vibrational

Figure 7.12   *Diagram illustrating electronic states in fluorescence*

For a molecule to fluoresce, it has to be excited in advance. The excitation spectrum of a molecule is identical with its UV absorbance spectrum. Fluorescence arises when some of the absorbed energy is converted to heat before the excited electron returns to its ground state (Figure 7.12). The energy not converted to heat is emitted as light of a longer wavelength than the excitation wavelength as it contains less energy.

7.8.3   Mass Spectrometry

Another spectroscopic technique called MS is also often used in bioanalysis. This technique is not based on interaction with light but requires the formation of ions in order to give a signal. The ions can be formed under different conditions and are then accelerated in an electrical field. Subsequently, the ions are analyzed in the mass separator due to their mass-to-charge ratio (m/z). The mass separator has to be kept at a high vacuum in order to avoid interaction with air molecules. When all ions have the charge +1, the m/z is numerically equal to the mass. If the charge is +2, the m/z is $1/2$ of the mass. MS is a highly valuable detector for bioanalysis because of its high sensitivity and selectivity. Peaks that are not resolved in chromatography can still be measured separately and used for quantitative analysis if their mass is different. A more detailed discussion of MS is given in the last part of this chapter.

7.8.4   UV Detection

The UV detection is based on the analytes' absorption of UV light (see Section 7.8.2). Analytes capable of absorbing UV radiation can be detected. This requires the analyte to contain a *chromophore*, being at least one double bond in the molecule. The wavelength range is from 190 to 400 nm, and at higher wavelengths the visible range up to about 800 nm can be used. Colored analytes are more selectively detected in the visible region, although

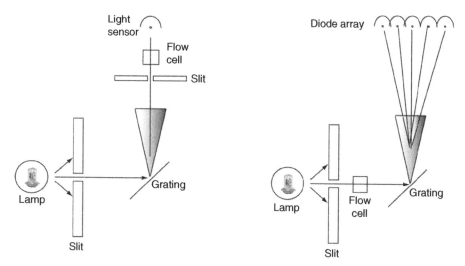

Figure 7.13   *Schematic diagram of a single wavelength UV detector (left) and a diode array detector (right)*

they also will absorb light in the UV region. It is first and foremost the high operational stability and ease that make the UV detector the preferred detector when sensitivity is no problem. It is also very good when using gradient elution. The lower limit of detection is adequate for some analyses, but it is not sufficiently sensitive to be used for analysis of low concentrations of drugs in biological material. Figure 7.13 shows a sketch of the UV detector.

The eluent from the column is directed to a flow cell through which UV light of a defined wavelength also is directed. The radiation that passes through the flow cell is detected by a light sensor.

It is common to use a deuterium lamp as a radiation source. A continuum of light is emitted in the whole UV range, and the detector can therefore be used, continuously variable, in the range of 200–400 nm. For optimal detection sensitivity of the substances, they should be measured at their maximum UV absorbance. A monochromator ensures that UV radiation of the correct wavelength is directed through the flow cell.

The measurement of absorbed radiation is, according to Beers' law, proportional to the concentration of the substance in the mobile phase, to the path length of radiation through the flow cell, and to the absorptivity of the substance.

Figure 7.14 shows a sketch of the flow cell. The eluent from the column flows through a Z-shaped channel in the cell. The UV radiation passes the flow cell through two quartz windows that do not absorb UV radiation. The path length of the flow cell is in the range from 6 to 60 mm, and the volume for a standard cell is in the order of 6–10 µl. When very efficient separation is obtained, such as in UHPLC, it can be necessary to use a flow cell with a smaller internal volume in order to avoid mixing of the separated peaks in the cell. A typical flow cell for UV detection in UHPLC has a path length of 10 mm and a volume of 0.5 µl.

In the diagram of the *single wavelength UV detector* shown in Figure 7.13, the monochromatic radiation of the selected wavelength passes through the flow cell and is directed

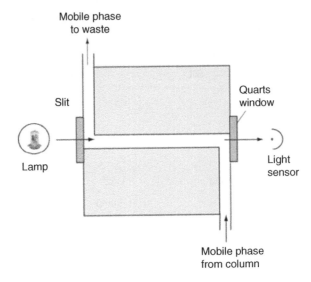

Figure 7.14   Sketch of flow cell in a UV detector

toward the detection unit. In another type of UV detector called a diode array detector (DAD), the polychromatic radiation is passed through the flow cell. After the flow cell, the transmitted light is split in an optical lattice into the individual wavelengths, and the intensity of each of these wavelengths is measured by a number of (array of) photodiodes. There may be up to several hundred diodes in series to measure the intensity of the array of UV radiation. *Diode array detectors* offer several possibilities. A full-UV spectrum of the peak can be recorded "on the fly," which can be used in the identification of the substance. It is also possible to choose a few selected wavelengths, so that each substance in a sample can be detected at the optimal wavelength.

7.8.5   Fluorescence Detection

All substances that are fluorescent also absorb radiation, but it is far from all substances that absorb radiation and also fluoresce. The fluorescence detector is therefore more selective than the UV detector. It also has a lower limit of detection than the UV detector, but this is only true for a selected number of analytes. For such compounds, it can be used to detect low concentrations of substances in biological samples.

Figure 7.15 shows a schematic diagram of the fluorescence detector. The lamp is usually a xenon lamp emitting an intense continuum of light in the whole UV region and in parts of the visible region. The fluorescence intensity is proportional to the energy input.

The wavelengths of the excitation and emission (fluorescence) radiation are controlled by monochromators. The molecules are excited using an intense light beam of a selected wavelength (the *excitation wavelength*). Some of the absorbed energy is released by heat dissipation, and the rest of the energy is emitted as light having a longer wavelength (the *emission wavelength*) than the excitation wavelength. The excitation wavelength can be chosen from the UV spectrum of the compound (Figure 7.16). Normally, it is preferred

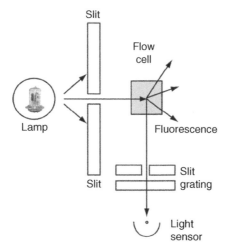

Figure 7.15   *Schematic diagram of a fluorescence detector*

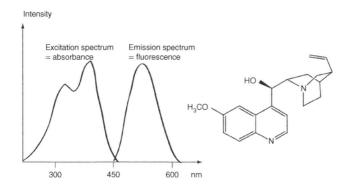

Figure 7.16   *Excitation and emission spectra of quinine*

to excite the compound at the wavelength where it has its maximum UV absorption, but for selectivity reasons another wavelength can be chosen. Also, the emission wavelength chosen for measuring can be varied, as the emission spectrum also covers a range of wavelengths.

For dilute solutions, the following equation expresses how the fluorescence is dependent on a number of parameters:

$$F = \Phi \cdot I_o \cdot a \cdot b \cdot c \tag{7.3}$$

where the fluorescence intensity, F, is proportional to the quantum yield, Φ; I_o is the intensity of the excitation radiation; a is the absorbance of the compound; b is the path length of the detector; and c is the concentration of the compound in the mobile phase. The quantum yield is a compound-dependent constant in a given environment.

Fluorescence is emitted in all directions but is normally measured at a 90° angle to the excitation radiation. This reduces background noise, and in principle the fluorescence is

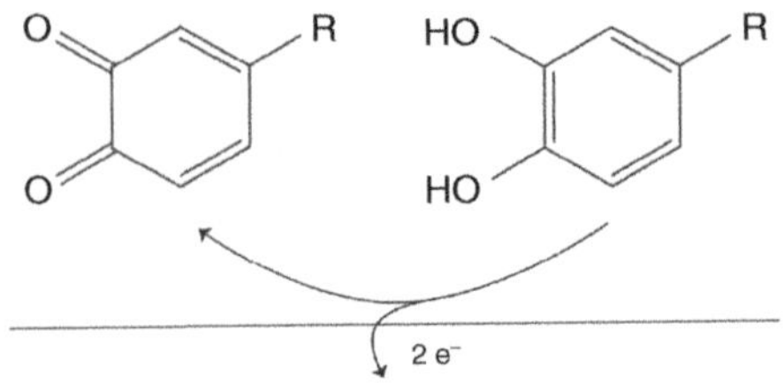

Figure 7.17 *Oxidation of catecholamines release electrons that are detected by the electro-chemical detector*

measured on a dark background. This is also why the fluorescence detector has a lower limit of detection than the UV detector. Because fluorescence intensity is proportional to the intensity of the radiation that excites the molecules of the substance, an increase in radiation intensity will provide a proportional increase in fluorescence intensity.

7.8.6 Electrochemical Detection

Electrochemical detection is a selective detection principle for compounds with electro-chemically active groups that can be either reduced or oxidized. Oxidation is in practice much easier to perform than reduction because the reduction mode requires that oxygen is totally removed from the mobile phase. The oxidation is performed at a given voltage, typically between +0.3 and +1.0 V. The higher the voltage, the more substances will be oxidized (Figure 7.17). The detector measures the current as a function of oxidation. The detector is especially well suited for detection of phenols, amines, and thiols.

The detector is not as robust as the UV and fluorescence detectors, and the electrodes can be contaminated by impurities in the mobile phase and in the samples. Pulsed amperometric detection expands the applicability of electrochemical detection to cover alcohols and carbohydrates. In this mode, carbohydrates can be detected with high sensitivity, but it has to be done at high pH (pH 12–13), and this of course puts some extra demands on the HPLC system.

7.8.7 Radiochemical Detection

Radiochemical detection is primarily used when identification of unknown metabolites is needed. The advantage is the very high degree of selectivity when the radiochemical detection is applied. When a peak in a chromatogram is also detected by the radiochemical detector, the peak represents either the compound labeled with the radioactive isotope or a transformation product thereof (a degradation product or a metabolite). The principle is therefore important when the metabolite profile of a drug substance is to be elucidated.

The most often used isotopes are ^{3}H and ^{14}C, but ^{35}S and ^{32}P also can be used. These isotopes emit β-radiation, which is absorbed by a scintillator and in this way is converted to photons. The photons are detected by a photomultiplier. On-line radiochemical detection therefore needs to have a scintillator liquid mixed into the effluent from the LC column. Another possibility is to perform fraction collection of the column effluent in a scintillation vial with scintillator liquid added, and then perform the measurement in a scintillation counter. This last principle can help to lower the limit of detection because the scintillation counting of each vial can be extended.

7.8.8   Combination of Detectors

Biofluids contain many compounds that may interfere with the analyte to be determined. It is therefore important to ensure that the chromatographic peak used for the quantification is the analyte in question and only this analyte. For this purpose, it is possible to combine detectors in series, and the combination of a UV detector, a fluorescence detector, and an electrochemical detector in series is a powerful tool. The detectors have to be coupled in the order given above. The fluorescence detector cell does not withstand higher back pressures, and the electrochemical detector has to be placed as the last one in the series as no back pressure is allowed. The major advantage of this setup is that the relative detector signals achieved for a given compound in the three detectors should be constant for a reference standard and for the corresponding peak in a complex sample. If this is not the case, there probably is some interference coming from other compounds in the sample, and this will bias the analytical result.

When also an MS detector is to be included in such a series, it is necessary to split the eluent after the UV detector, with a part going to the MS and the other part to the fluorescence and electrochemical detectors.

7.9   Mass Spectrometric Detection[1]

Mass spectrometers coupled to an LC system have become a very common tool in bioanalysis. Not only are mass spectrometers well suited to perform identification of new metabolites in biological matrices, but also they are used to confirm the identity of compounds as well as to quantify these in very low concentrations (even down to femtomolar levels). Most drug substances and their metabolites can be determined using MS. It is important, however, that the compounds to be analyzed can be ionized and can be transferred to the gas phase. *To compare with UV detection*: to be able to determine compounds using UV, the analytes need to absorb UV light at the wavelength at which one wants to run the analysis.

Figure 7.18 shows a typical setup of an LC-MS system. The coupling between HPLC and the mass spectrometer is challenging. In this coupling, or interface, the analytes are ionized and transferred to the gas phase. After this, the mass analyzer separates the ions based on their mass-to-charge ratio (m/z ratio; m is the exact mass of the analyte, and z is the number of charges of the analyte). There are several interfaces and mass analyzers used in bioanalysis, each with their specific features and specific optimal applications.

To get insight regarding which choices to make when running a bioanalytical experiment on the LC-MS, this chapter will discuss:

- The most common interfaces used in bioanalysis
- The most common mass analyzers

---

[1] This section (7.9) and its figures are based on "Chromatography—Basic Principles, Sample Preparations, and Related Methods," by Lundanes, Reubsaet, and Greibrokk. Pages 85–95. 2014. Copyright Wiley-VCH Verlag GmbH & Co. KGaA. Reproduced with permission of Wiley-VCH Verlag GmbH & Co. KGaA.

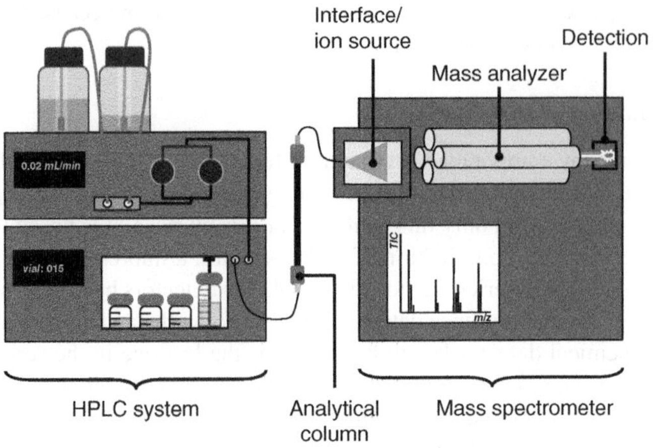

Figure 7.18 *Typical setup of a liquid chromatography–mass spectrometer (LC-MS) system*

- In which mode the instrument needs to be run to obtain the information needed
- How to interpret the main features of the mass spectrum
- Which pitfalls might occur
- Common definitions.

Box 7.1 Calculation of the Mass of a Compound

Because mass spectrometers can have a high mass accuracy, it is of major importance that one uses the right mass values to calculate the mass of a compound. There are mainly four definitions for the mass of a molecule: the molecular mass, the exact mass, the monoisotopic mass, and the nominal mass. These resemble each other but cannot be interchanged.

Molecular Mass

This is the average mass of a compound; it is the mass to be used in molar calculations and can be found by calculating the sum of atomic masses of each element multiplied by its presence in the molecular formula. Using chlorpromazine ($C_{17}H_{19}ClN_2S$) as an example, the molecular mass is $17 \times 12.011 + 19 \times 1.008 + 1 \times 35.453 + 2 \times 14.007 + 1 \times 32.065 = 318.871 \text{g/mol}$.

Nominal Mass

This mass is obtained by calculating the sum of the mass numbers of the most abundant isotopes in a molecule. As an example: chlorpromazine ($C_{17}H_{19}ClN_2S$): nominal mass is $17 \times 12 + 19 \times 1 + 1 \times 35 + 2 \times 14 + 1 \times 32 = 318$.

Monoisotopic Mass

This is the exact mass of the molecule calculated from the sum of the most abundant isotopes. For example, for chlorpromazine ($^{12}C_{17}{}^1H_{19}{}^{35}Cl^{14}N_2{}^{32}S$), the molecular mass is $17 \times 12.000 + 19 \times 1.008 + 1 \times 34.969 + 2 \times 14.003 + 1 \times 31.972 = 318.099$.

Exact Mass

This is the exact mass of a certain isotope species of the molecule. This can be the monoisotopic mass but also the mass of the molecule containing various isotopes of the same element. This mass can be found by calculating the sum of certain isotopes in the molecule. Example: chlorpromazine containing one ^{37}Cl isotope: ($^{12}C_{17}{}^1H_{19}{}^{37}Cl^{14}N_2{}^{32}S$): exact mass is $17 \times 12.000 + 19 \times 1.008 + 1 \times 36.966 + 2 \times 14.003 + 1 \times 31.972 = 320.096$.

Both monoisotopic mass and exact mass are used in MS.

7.9.1 Mobile Phases in LC-MS

When a mass spectrometer is used as detection in HPLC, there are some restrictions when it comes to mobile phase components. As MS operates such that everything that elutes from the column is transferred into the gas phase (both analytes *and* mobile phase), it is of principal importance that volatile components are used as mobile phase constituents. It should also be noted that the used components are of a suitable purity grade (little residue). This is often referred to as *LC-MS grade*. This is unproblematic for often-used solvents like water, acetonitrile, methanol, and THF. Because only a few buffer components are volatile, the choice of pH in the mobile phase is limited. The following volatile reagents can be used: formic acid, acetic acid, ammonia, ammonium formate, and ammonium acetate. Buffers should be prepared from a combination of these.

In other words, when an acetate buffer pH 4.8 needs to be prepared, the constituents should be acetic acid and ammonium acetate and not acetic acid and sodium acetate. Although both are acetate buffers, the latter consists of nonvolatile sodium and is, as such, not suitable for LC-MS purposes.

7.9.2 Interfaces and Ionization Methods

There are numerous ionization methods that allow the formation of ions to carry out MS; however, in this chapter we only focus on those most common in HPLC-MS. The challenge in coupling HPLC to MS is that the chromatography operates with liquids and under high pressure while the detector is under a (usually high) vacuum. In MS, the link between the chromatography and the mass spectrometer is called the interface. Here, ionization and transition from the liquid phase to the gas phase of the compounds occur.

In LC-MS-based bioanalysis, the following interfaces are most used: electrospray ionization (ESI), atmospheric pressure chemical ionization (APCI), atmospheric pressure

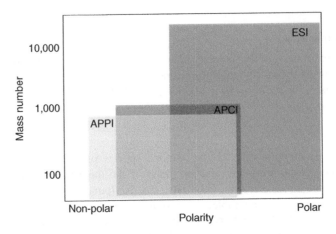

Figure 7.19 *Choosing the right interface depends on the analyte size and the polarity*

photo-ionization (APPI), and inductively coupled plasma (ICP). Although there are several more interfaces existing, focus will be on the ones mentioned above.

ICP mostly is used in the determination of metals (elemental analysis) and has its specific applications.

ESI, APCI, and APPI are related ionization techniques with a wide range of applications. Choosing the suitable interface depends on the nature of the analytes to be analyzed. In general, polarity and size determine which of the three interfaces to choose (see Figure 7.19).

ESI is used when polar compounds (neutrals, acids, and bases) need to be ionized. The electrospray itself is operated at atmospheric pressure. A schematic representation of the process of electrospray is shown in Figure 7.20: the eluate from the column (containing mobile phase and analytes) enters a capillary where a high voltage is applied, typically +5 or −5 kV. At the outlet of the capillary, a nebulizing gas (mostly N_2) is mixed with the liquid flow to facilitate formation of droplets. As can be seen from Figure 7.20, drying gas (desolvation gas) also is introduced. This is performed in the opposite direction of the flow.

Droplets leaving the capillary are highly charged due to ion accumulation caused by the high voltage. On its way to the entrance of the mass spectrometer, the highly charged droplets decrease in size because the mobile phase evaporates. This causes an increase of intrinsic charge repulsion between the ions with the same charge. When the repulsive forces inside the drop exceed the surface tension (Rayleigh limit), the droplet explodes in smaller droplets (Coulomb fission). These droplets undergo the same process again (size reduction through desolvation, followed by Coulomb fission due to the high degree of charge repulsion). This is a repetitive process yielding ions in the gas phase: either through ion evaporation (ion evaporates from the droplet) or as a charge residue (after the solvent is evaporated, the ions are left over).

Depending on the polarity of the initial voltage applied, the mass spectrometer is operated in the positive mode or the negative mode. In the positive mode (the polarity of the voltage at the outlet of the electrospray capillary is positive), protonated ions are detected. In the negative mode (the polarity of the voltage at the outlet of the electrospray capillary is negative), deprotonated ions are detected. The detected ions are called pseudo-molecular

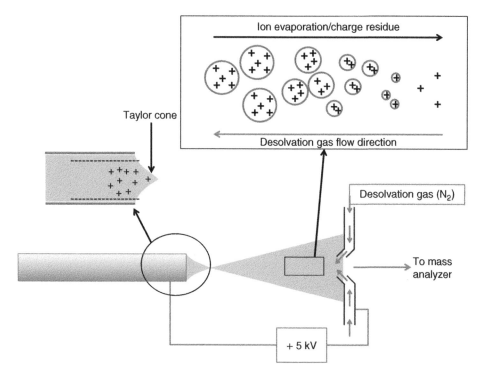

Figure 7.20 *Schematic representation of the electrospray and the formation of ions*

ions (for simplicity, called molecular ions throughout this chapter): $[M + nH]^{n+}$ for positive ions and $[M - nH]n^-$ for negative ions (M is the mass of the compound, H is the mass of the proton, and n is the number protons accepted or donated). In most cases, $n = 1$, yielding a molecular ion with charge +1 or −1 ($z = 1$) (see Box 7.1).

Neutral polar compounds either accept or donate protons under given conditions, yielding positive ions or negative ions either in the mobile phase or during the ESI process. The actual ionization process for acids and bases in ESI occurs in the mobile phase by pH adjustment. The basic compound haloperidol (pK_a 8.8) is protonated and thus positively charged (Figure 7.21a) at pH 6, whereas the acid compound acetylsalicylic acid (pK_a 3.5) is negatively charged under these conditions (Figure 7.21b).

This means that a molecule like haloperidol (Figure 7.21a) is measured at an m/z value of $[M + H]^+/z = (375.14 + 1.00)/1 = 376.14$, and acetylsalicylic acid (Figure 7.21b) at an m/z value of: $[M - H]^-/z = (180.04 - 1.00)/1 = 179.04$. In cases where more charges are involved, the measured m/z is obtained in the same way: the peptide Asp–Arg–Val–Tyr–Ile–His–Pro–Phe (angiotensin II) with a mono-isotopic mass of 1045.53 can be triply charged (three protons are added): this yields an m/z value of 349.51: $(1045.53 + 3.00)/3$. It has also been detected as doubly charged (two protons added): this yields an m/z value of 523.77 : $(1045.53 + 2.00)/2$ (see Figure 7.22).

The optimal flow for ESI is below 50 µl/min. High flow rates not only hamper the evaporation of ions but also dilute the analyte, thus leading to bad sensitivity. In other words, the

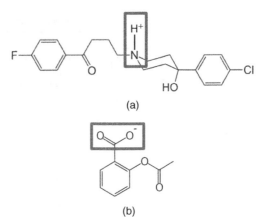

(a)

(b)

Figure 7.21   (a) Protonated haloperidol (thus positively charged) having [M + H]⁺ = 376.14; and (b) de-protonated acetylsalicylic acid (thus negatively charged) having [M − H]⁻ = 179.04

Figure 7.22   *Angiotensin II consists of eight amino acids, and three of them can be positively charged (gray boxes). The mass of angiotensin II is 1045.53; adding three protons makes up a mass of 1048.53 and will be detected like* [M + 3H]³⁺ = 349.51. *Adding two protons makes up a mass of 1046.53 and will be detected like* [M + 2H]²⁺ = 523.77

lower the flow rate, the better the performance of the electrospray. An ESI can be operated at flow rates down to the nl/min rate. In this case, the ESI is termed nanospray. Although not always as easy to operate routinely, the performance in terms of sensitivity of a good working nanospray is much better than that of conventional ESI. In combination with conventional LC, flow rates in the range of 0.2–0.5 ml/min are often used, although these rates are nonoptimal for the ESI.

APCI mainly is used for small relatively nonpolar compounds. This does not mean that polar compounds not can be ionized by this technique (although ESI would be the first choice in such case). The difference with ESI is that ionization in APCI takes place in the ion source. As can be seen from Figure 7.23, a needle is situated near the outlet of a heated capillary. Through a high potential applied on this needle (corona discharge needle), a plasma of ions is created around it. This causes ionization, under atmospheric pressure, of gas molecules like N_2 to form $N_2^{+\bullet}$ and $N_4^{+\bullet}$. N_2 originates from the sheath gas, which is used in the APCI process. These reactive $N_2^{+\bullet}$ and $N_4^{+\bullet}$ produce secondary ions from the evaporated solvents (H_2O). After a series of reactions, H_3O^+ and $OH^\bullet$ are produced, which in their turn protonate and deprotonate the substances to form, respectively, $[M + H]^+$ and $[M − H]^-$. Thus, the result is the same as with ESI: protons are accepted or donated, yielding positively charged molecular ions or negatively charged molecular ions.

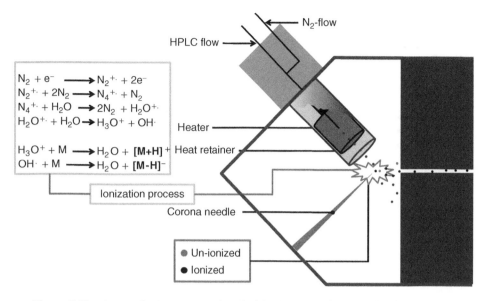

Figure 7.23 *Atmospheric pressure chemical ionization with gas molecular reactions*

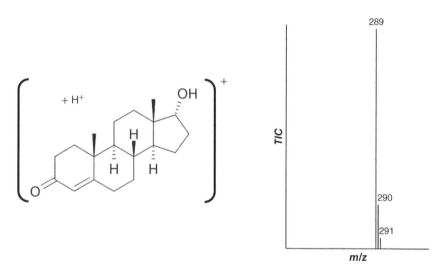

Figure 7.24 *Protonated epitestosterone gives rise to m/z 289 (= [M + H]⁺) in the mass spectrum (m/z 290 and 291 are isotopes)*

Figure 7.24 shows the mass spectrum of the nonpolar epitestosterone measured by MS using an APCI interface. Epitestosterone does not contain basic or acidic functional groups, but it yields a positive molecular ion.

Another difference with ESI is the mechanism of getting mobile phase and analytes in the gas phase. In APCI, the mobile phase enters the interface through a capillary, which is heated up to typically 400–500 °C. Everything that goes through the capillary is evaporated

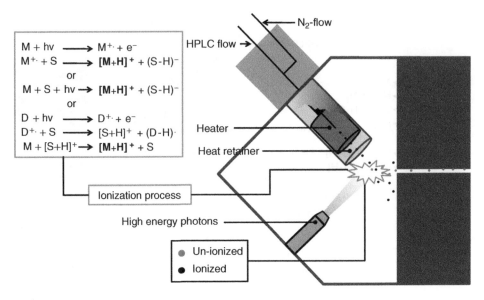

Figure 7.25   *Atmospheric pressure photo-ionization with the three types of ionization*

instantly and at the outlet mixed with a sheath gas (usually N_2). One should bear in mind that the conditions used in APCI might lead to degradation of thermolabile compounds. *In-source hydrolysis* of esters will lead to degradation products, which might be detected at a different m/z ratio in the same spectrum. APCI can be used for those applications where flow rates can be high. The optimal flow rate for APCI varies between 0.2 and 2 ml/min.

APPI is closely related to APCI and is also mainly used for small relative nonpolar compounds. The sample introduction is similar to that of APCI: mobile phase and analytes are vaporized in a vaporizer tube at high temperature (400–500 °C). As can be seen from Figure 7.25, the geometry of the ion source is similar to that of the APCI; however, instead of the corona discharge needle, an UV lamp is used: ionization in APPI is caused by photons. The energy of these photons is transferred directly to the analyte (causing *direct ionization*), to the solvent (causing *solvent-assisted ionization*), or (if used) to a dopant (causing *dopant-assisted ionization*). In all cases, molecular ions are formed.

Direct ionization is caused by the energy transfer to the analyte, yielding a radical cation. This radical then obtains a hydrogen originating from solvents like water or methanol. Excitation of solvent molecules directly also occurs. This *solvent-assisted ionization* also leads to transfer of a proton from the solvent to the analyte. *Dopant-assisted ionization* is used to increase the ionization efficiency of the analytes. For this purpose, so-called dopant molecules (like toluene) are used. These are ionized by the high-energy UV source to radical cations, which in turn produce molecular ions. Formation of positive molecular ions and negative molecular ions occurs via different mechanisms, but these are not discussed here because this is outside the scope of this book.

ICP is an ion source that is used in elemental-specific analysis. It is commonly used to analyze metals, but it is also applied to nonmetal analysis. ICP is used in both

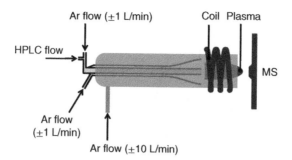

Figure 7.26 *The high velocity of electrons (initiated by a spark) caused by a frequency of 30–40 MHz of the coil causes argon gas to ionize. In this way, a plasma at high temperature is created*

pharmacokinetic studies of metallodrugs like oxaliplatin and cisplatin as well as to perform toxicological analysis.

In ICP, all compounds entering the interface are atomized because of an extremely high temperature. This temperature can vary between 7000 and 10 000 K and is generated by the ICP torch. This torch consists of three tubes and functions as follows: argon gas flows into the ICP torch, and through a spark electrons are generated and accelerated within an electromagnetic coil. Collision of the high-velocity electrons with argon forces ionization and production of a plasma. The metal (or nonmetal) atoms are ionized simply through the loss of an electron, producing singly positive charged species (Figure 7.26).

Typical spectra generated after ICP ionization only consist of the single element with its isotopes. The *m/z* value detected is the mass of the element (M^+); this is in contradiction to the molecular ions produced by ESI, APCI, and APPI, which produce protonated or deprotonated ionic species that differ in the single charge state one mass unit from their unionized form. Typical elements that can be analyzed using ICP are Cd, Cr, Mo, Se, Zn, Cu, Fe, Pt, Pb, Co, Al, and Mg.

7.9.3   Parameters Characterizing the Mass Analyzers

As ions leave the ion source, they enter the mass analyzer. The mass analyzer is operated under (high) vacuum and allows ions to be separated from each other on the basis of their *m/z* value. Depending on the type of mass analyzer (discussed further in this section), the performance can vary. The performance of the mass analyzer is described, among others, by the following parameters: mass resolution, mass accuracy, and scan speed.

Mass resolution (R) describes the ability to separate *m/z* values from each other. The higher the value for R, the better it separates closely related *m/z* values.

Mass resolution can be calculated using the following equation:

$$R = m/\Delta m \tag{7.4}$$

where *m* is the *m/z* value of a measured signal; and Δm is the width of this signal at half peak height. Values for R can vary between 1000 (bad resolution) and 1 000 000 (very good resolution). The impact of a good resolution on the mass spectrometric result is obvious

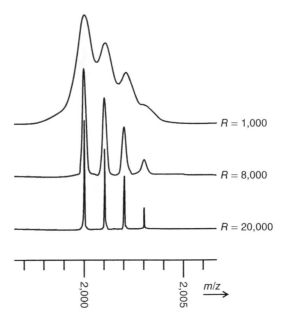

R = 1,000

R = 8,000

R = 20,000

2,000

2,005

m/z

Figure 7.27   *Three mass spectra with increasing mass resolution*

from Figure 7.27. Here, three mass spectra from the same ion are shown with an increasing mass resolution. The higher the mass resolution, the more information is revealed.

Mass accuracy (*E*) describes the difference between the measured *m/z* and the theoretical *m/z* value. The lower the value for E, the more correct is the measured *m/z* value. Mass accuracy can be calculated using the following equation:

$$E = \left(\frac{\delta m}{m_{measured}} \right) \times 10^6 \, \text{ppm} \tag{7.5}$$

Here, $m_{measured}$ is the *m/z* value of the measured signal; and δm is the difference between the theoretical *m/z* of the analyte and the measured *m/z* value. Mass accuracy is expressed in parts per million (ppm). Values can be as low as 1–2 ppm (very good accuracy).

Scan speed is determined by the scan cycle time. This cycle time is the time needed to obtain the mass spectrum. The scan speed is given in hertz. Depending on the application and the mass analyzer used, scan speeds can be up to several hundred hertz but also down to a single scan per second. A known challenge with low scan speeds is their use in chromatographic separations where the chromatographic peak is very narrow (e.g., in UHPLC). As a rule of thumb, one should have at least 12–15 data points per peak. This implies that narrow peaks need high-scan-speed mass spectrometers.

7.9.4   Commonly Used Mass Analyzers in Bioanalytical LC-MS

A *quadrupole mass analyzer* is a mass filter that filters ions on the basis of their *m/z* value. The quadruple consists of four identical rods that are placed parallel to each other. Both opposite pairs of rods are connected electrically. By applying both a certain direct current

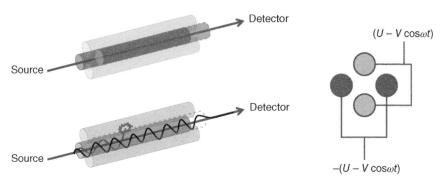

Figure 7.28 (a) Ions move from the source toward the detector in the z-direction. Only the stable oscillating ions will reach the detector (black line). Unstable ions will collide with one of the quadrupoles (gray line). (b) x/y view of the quadrupole. The electrical field applied is composed of a DC component (U) and an RF component (Vcos ϖt)

(DC) and a radio frequency (RF) on one of the pairs (U-Vcos ϖt) and the opposite DC and RF on the other pair −(U-Vcos ϖt), an oscillating electrical field is created (Figure 7.28). Ions enter this oscillating electric field in the z-direction and fly toward the detector. Due to the oscillating electrical field, ions will, in addition to the z-movement, also start to oscillate in the x- and y-directions. The flight trajectory is thus not a straight line from entrance to exit of the quadrupole. Those ions that have a stable trajectory will reach the detector. Those ions that have an unstable trajectory will collide with one of the quadrupoles and will not be detected.

Only certain combinations of DC and RF will allow a particular ion to pass the quadrupole filter through a stable trajectory. Each particular m/z value has certain combinations of DC and RF at which they oscillate stably. The quadrupole is calibrated such that DC and RF combinations are correlated to m/z values. By varying DC and RF values in a controlled way, whole mass spectra can be obtained.

A single quadrupole mass spectrometer is operated as described above. This system contains only one quadrupole as mass analyzer. It is suited for both full scan analyses and selected ion monitoring (SIM) analyses. Fragmentation in this mass analyzer is not possible. A single quadrupole MS has unit resolution (R < 1000); moderate accuracy (E > 100 ppm) at a high scan speed; and, when operated in the most sensitive way (SIM; discussed in Section 7.9.6), picogram sensitivity.

A triple quadrupole functions in principle the same as a single quadrupole but has the advantage of performing fragmentation within the analyzer. A triple quadrupole consists of two quadrupole mass analyzers and a collision cell. Figure 7.29 shows what a triple quadrupole looks like. A triple quadrupole can be operated as a single MS or as a tandem MS. When it is operated as a single MS, no fragmentation is carried out and either the first or the last quadrupole is used as mass analyzer. However, this mode is usually not used in a triple quadrupole. A triple quadrupole is mainly used as tandem MS.

Fragmentation in a triple quadrupole is performed as follows: in the first quadrupole, the ions to be fragmented are filtered. These ions are transferred to the collision cell. This can be a quadrupole (hence the name) or a hexapole that contains an inert gas (usually Ar). By accelerating the ions in the z-direction toward the exit of the collision cell, ions

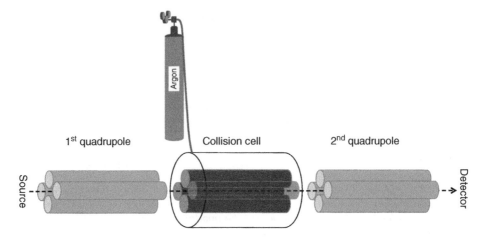

Figure 7.29 *Schematic representation of a triple quadrupole mass spectrometer*

collide with the gas molecules present and fragment into lesser ions (or *neutrals*). This is called *collision-induced dissociation* (CID). The fragment ions are transferred to the last quadrupole, which is then operated as a mass analyzer.

An *ion trap mass analyzer* can be operated in the single MS mode and in the fragmentation MS mode. Mass separation in the ion trap is based on stabile trajectories of ions in an oscillating electrical field. There are two ion trap configurations: the 3D ion trap (or Paul trap; see Figure 7.30) and the 2D ion trap (or linear ion trap; see Figure 7.31). The principle of the ion trap is explained using the classical 3D ion trap. It consists of a ring electrode and two end-cap electrodes on which AC and DC voltages are applied. Ions enter the ion trap with high velocity through a hole in one of the end-cap electrodes. The presence of helium in the ion trap (present at low pressure) causes collisions that slow down the ions to such an extent that they get trapped in the electrical field. The oscillating electric field makes the ions move within a certain space inside of the ion trap. The collisions with He are mild and will usually not cause any fragmentation.

After this trapping process, ions are destabilized in the trap by applying an increasing voltage on the electrodes, making the ions move in an increasing area within the trap. As the voltage becomes high enough, the ions will leave the trap (light ones first, heavy ones last), allowing detection. This is typically performed in the single MS mode: all ions that leave the ion trap are detected as molecular ions. When operated in the single MS mode, the collision with the existing helium atoms is so mild that no fragmentation occurs.

When the ion trap is operated in the fragmentation MS mode, the ions are trapped, fragmented, and scanned *in the same mass analyzer*. The ion trap is then operated such that after the trapping process, all ions that are not of interest are destabilized such that only the *m/z* value to be fragmented is left in the analyzer. This particular ion is then accelerated in the ion trap to such an extent that it will not leave the trap, but has enough velocity to collide with the helium that is present and form fragments. After this process is carried out, all fragments are trapped in the center of the ion trap and in the last step analyzed.

In a *time-of-flight* (ToF) *mass analyzer*, the *m/z* value of an ion is determined by its flight time in a field-free tube. Ions produced in the ion source enter the flight tube through pulsed

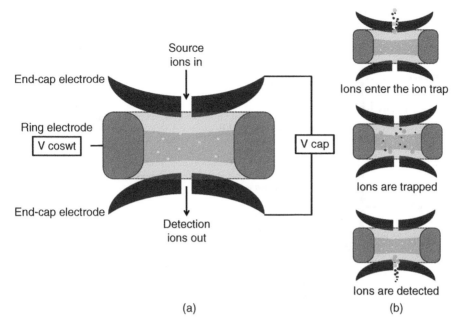

Source
ions in

End-cap electrode

Ring electrode

V coswt

V cap

Ions enter the ion trap

Ions are trapped

End-cap electrode

Detection
ions out

Ions are detected

(a)   (b)

Figure 7.30   *The ion trap consists of a ring electrode and two end-cap electrodes (a). On the right (b), from top to bottom: ions enter the ion trap and are cooled down by He (white dots). After the ions are trapped, they are scanned out of the ion trap. Light ions leave the trap before the heavier ones*

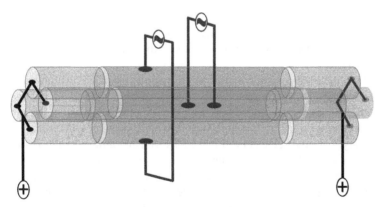

Figure 7.31   *The linear ion trap consists of three quadrupole-like electrodes where the first and last quadrupoles are coupled to a direct voltage with the same polarity. An ion trap field is created within the second quadrupole*

extraction. This extraction is based on a pulsed electrical field between two plates (a repeller plate and extraction grid). The ions will be accelerated in this field and leave the extraction grid with a certain kinetic energy, which makes them move toward the detector. As there is no electrical field between extraction grid and detection (field-free tube), the ions will keep their velocity at which they entered the flight tube. The m/z value is inversely proportional

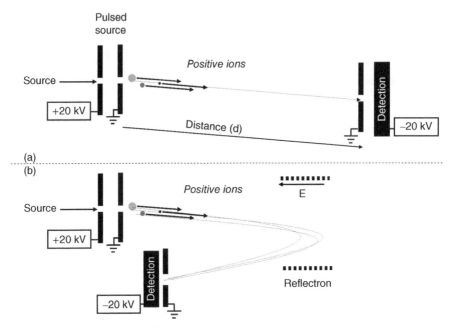

Figure 7.32 *(a) Linear time-of-flight mass analyzer. Heavy ions (largest circles) travel slower than lighter ions (smallest circles). The velocity of the ions traveling through the field-free flight tube thus correlates with the m/z value. (b) In a reflectron ToF mass analyzer, all ions are deflected by a reflectron to reach higher resolution and mass accuracy by a longer flight path*

with the travel time: heavy ions have a longer flight time than light ones. The travel time is calculated from the pulse to the impact on the detector. Figure 7.32 shows the concept of the ToF analyzer.

Figure 7.32a shows the principle of ToF MS through a linear ToF. Although a ToF mass analyzer can function in this geometry, it has poor resolution and accuracy. This is caused by the fact that the kinetic energy of ions with a certain *m/z* value is not the same after a pulsed electrical field is applied. In other words, there is a spread of kinetic energy (and thus a somewhat varying initial velocity) within the ions with the same *m/z* value (isobaric ions). This will cause varying flight times within the population of isobaric ions. A reflectron ToF (see Figure 7.32b) is designed to correct for this. A reflectron is a set of metal plates on which a voltage is applied. In this way, an electrical field is created that repels the ions entering it. The spread in kinetic energy is minimized because of this reflection causing a minimal difference in velocity within an isobaric ion population. The reflected ions enter a second flight tube and are detected at the end. Compared to a linear ToF mass analyzer, a reflectron ToF mass analyzer has very high resolution and mass accuracy.

Both linear ToF and reflectron ToF mass analyzers are used for single MS purposes. The upper limit of the *m/z* value that can be determined using ToF MS is 500 kDa or even higher. ToF mass analyzers are often coupled to other mass analyzers for fragmentation purposes. Known examples are Q-ToF and IT-ToF instruments. In the Q-ToF, the quadrupole is used for initial mass filtering as the first step in the fragmentation process. Filtered ions are then fragmented in a collision cell that is placed between the quadrupole and the ToF. The ToF is used to analyze the outcome of the fragmentation. In the IT-ToF, both mass filtering and fragmentation occur in the ion trap (as described here). The ToF is also in this type of instrument used to analyze the outcome of the fragmentation.

Ion mobility spectrometry–mass spectrometry (IMS-MS) resembles a ToF mass analyzer in the way that ions are separated in a drift tube. However, in contradiction to ToF, the IMS drift tube, or IMS mobility cell, contains an inert buffer gas (also called *carrier gas*) at a certain pressure and an electric field. With a velocity based on the shape, size, and charge, ions will move from inlet to detection. The principle of mass separation in IMS is as follows. Ions enter the drift tube and are attracted and accelerated to the exit because of the electrical field applied in the tube. This electrical field is created by a set of ring electrodes. In their movement from entrance to exit, the ions will collide with the inert buffer gas present, which flows into the opposite direction. This causes a drag force, which acts against the acceleration caused by the electrical field. The larger the cross-section of the ions with which the gas molecules collide, the lower the velocity of the ion. In other words, two ions with the same *m/z* value but with different sizes will be separated in the ion mobility cell. Because separation in the drift tube is based on shape, size, and charge, insight in the structure of the analyte can be obtained. Figure 7.33 shows the IMS mobility cell.

There are several modes of IMS-MS: the drift-time ion mobility spectrometry (DTIMS), high field asymmetric ion mobility spectrometry (FAIMS), and traveling wave ion mobility spectrometry (TWIMS) methods. It is not within the scope of this book to discuss these specific modes in detail.

The use of *Fourier transform* (FT) in MS is increasing due to the availability of user-friendly analyzers. There are two types of FT-based mass analyzers that are in use: ion cyclotron resonance (ICR) and orbitrap (OT). Although the mode of action of both systems is slightly different (it is not within the scope of this book to explain these mass analyzers in detail), both use the same way of determining the *m/z* value of an ion.

After ions are produced in the ion source and guided to the analyzer, they are trapped, either in a Penning trap (based on a magnetic field and an electric field for the ICR analyzer) or in an OT (based on an electrical field for the OT analyzer). The trapped ions move in an orbit in the trap, generating a charge and therewith an image current that can be measured on the outside of the trap. This image current is sinusoid, and the frequency of this sinusoid is reciprocal proportional to the *m/z* value. In other words, every *m/z* value generates a specific frequency. The lower the frequency measured, the higher the *m/z* value.

Because there are usually several ions present during a measurement, complex frequencies are measured, generated by several ions.

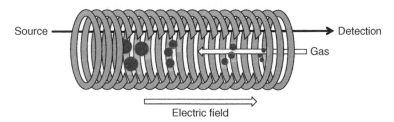

Figure 7.33 *Schematic representation of ion mobility spectrometry. The drift tube consists of several focusing rings, creating an electric field that is opposite in direction to the drift gas flow*

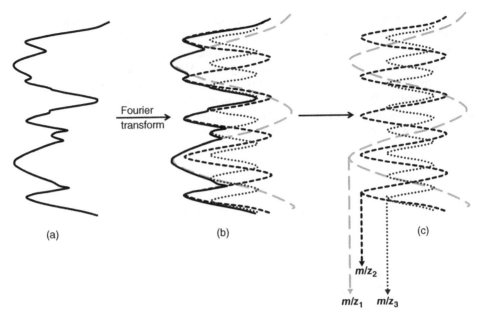

Figure 7.34   *(a) Electrical signal measured in ion cyclotron resonance or an orbitrap mass analyzer. (b) Signal is decomposed in all the frequencies. (c) Separate frequencies of three masses, where $m/z_3 < m/z_2 < m/z_1$*

To be able to extract the information needed from this complex result, an FT (see Figure 7.34) of the data is needed to be able to find the individual frequencies the signal is made up of, thus enabling us to find the separate *m/z* values. FT-MS is associated with very high mass resolution and mass accuracy.

7.9.5   Ion Detection

After ions have been separated in the mass analyzer, they are detected. Except for the FT mass analyzers, detection is based on the impact of an ion on a surface, generating a measurable current. There are several detectors used in MS that will be discussed briefly. One should keep in mind that the detector is only there to measure the presence of an ion. The combination with the mass analyzer allows researchers to perform mass measurement.

An *electron multiplier* is based on secondary emission, which means that impact of an ion on an emissive material causes the release of several electrons. These electrons will, in their turn, release several electrons per impact from the emissive material, thus amplifying the signal. Figure 7.35a shows a schematic representation of the electron multiplier.

A *microchannel plate* is also based on secondary emission. As can be seen from Figure 7.35b, an ion enters the microchannel and its signal is amplified (up to 10^5).

A *Faraday cup* is a detector based on the impact of an ion on a metal cup. After this, the ion's charge is neutralized, generating a current. This current is amplified and detected. The currents measured are proportional to the amount of ions that hit the metal cup. Figure 7.35c shows a schematic representation of the Faraday cup.

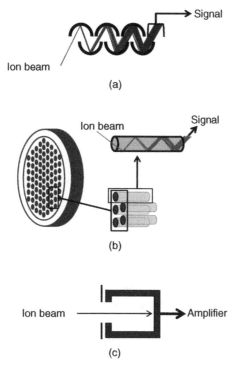

Figure 7.35 *Detectors used in mass spectrometry: (a) electron multiplier, (b) microchannel plate, and (c) Faraday cup*

In the mass analyzers described, usually the molecular ion ($[M + nH]^{n+}$) is measured. This is, in many applications, specific enough to perform both qualitative and quantitative analysis. However, in the case where isobaric compounds need to be separated or where low quantification levels are needed, *tandem* MS is carried out.

Separation of isobaric analytes is not unusual in drug analysis, where also metabolites of a specific drug are monitored. An example is shown in Box 7.2.

Box 7.2 Analysis of Ketobemidone and Two of Its Metabolites

Figure 7.36 shows the structures and the masses of ketobemidone and two of its metabolites.

From the masses of the analytes, it can be seen that it is challenging to separate ketobemidone, dihydroxyketobemidone, and p-hydroxymethoxynorketobemidone from each other. The latter two have the same mass. Fragmentation of such substances in the mass spectrometer yields fragmentation spectra that are specific for each of the analytes, thus allowing identification and differentiation between the metabolites, even if the precursor masses are identical.

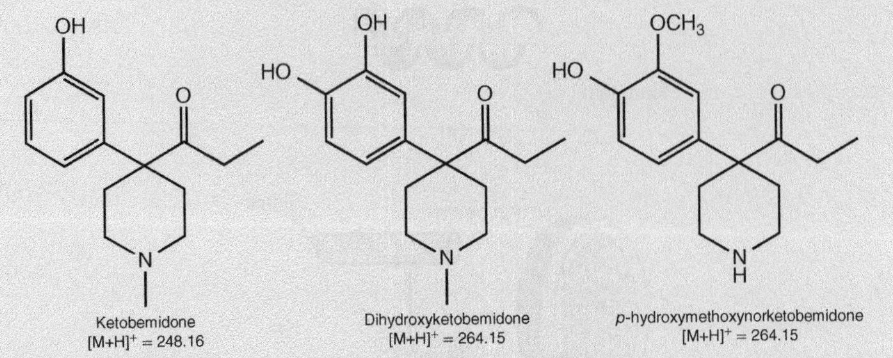

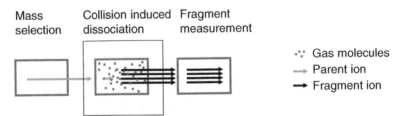

Figure 7.36   *Chemical structure and masses of ketobemidone and two of its metabolites*

Low quantification levels are obtained using tandem MS because the detector is operated in a more selective manner, thus lowering the noise of interferences. Due to the fragmentation, selective fragment ions arise that can be monitored. This topic is described in more detail in Section 7.9.6.

Fragmentation in tandem MS usually is carried out via CID. The general principle is as follows: in step 1, the ion of interest is selected. In step 2, when the impact is high enough that the molecular ion will fragment, this ion is caused to collide with the gas molecules present. These fragments are then detected in step 3. Figure 7.37 shows this general principle.

There are several ways to perform fragmentation depending on the geometry of the mass spectrometer. Examples of such systems are: triple quadrupole MS, ion trap, ion trap–ToF, ToF–ToF, ion trap–orbitrap, and quadrupole–orbitrap. Depending on this geometry, fragmentation occurs in *space* or in *time*.

Fragmentation in space is performed by those instruments that geometrically have separated step 1, step 2, and step 3. In other words, ion selection, ion fragmentation, and ion fragment detection are carried out in different sections of the instruments. A typical instrument that carries out fragmentation in space is the triple quadrupole mass analyzer: ion selection in the first quadrupole, ion fragmentation in the collision cell, and ion fragment analysis in the last quadrupole.

Mass selection   Collision induced dissociation   Fragment measurement

∴ Gas molecules
→ Parent ion
⟶ Fragment ion

Figure 7.37   *General principle of tandem mass spectrometry. Step 1: mass selection; step 2: collision-induced dissociation; and step 3: fragment measurement*

Fragmentation in time is performed by those instruments that have integrated steps 1–3. This means that ion selection, ion fragmentation, and ion fragment detection are performed in the same analyzer. A typical instrument that carries out fragmentation in time is the ion trap mass analyzer: The ion of interest is trapped in the ion trap as described earlier. Other ions that are trapped at the same time are subsequently destabilized such that only the ion of interest is left in the ion trap. In the second step, fragmentation of the trapped ion is carried out. The fragment ions are determined in the third step. In Box 7.3, the specific fragmentation pattern of peptides is described.

Box 7.3 Fragmentation of Peptides

Fragmentation mainly occurs in the peptide backbone between the amine nitrogen and α-carbon, between the α-carbon and the carboxylic acid carbon, or between the carboxylic acid carbon and the amine nitrogen. This yields so-called x_n-, y_n-, and z_n-fragments and a_n-, b_n-, and c_n-fragments. The x_n-, y_n-, and z_n fragments are to be read from the fragmentation site to the C-terminal (the charge remains on the C-terminal fragment). The a_n-, b_n-, and c_n-fragments are to be read from the N-terminal to the fragmentation site (the charge remains on the N-terminal fragment). The subscript n gives information on the number of amino acid residues. (See Figure 7.38.)

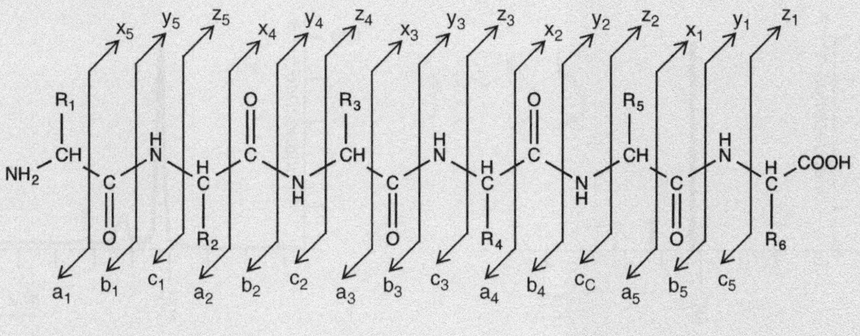

Figure 7.38 *Fragmentation possibilities of a hexapeptide and their fragment names*

7.9.6   How to Obtain the Information Needed and How to Interpret the Mass Spectrum

The LC-MS analysis mainly provides two results: the chromatogram where all substances are represented as chromatographic peaks, and the mass spectra that contains mass-related information of the analyzed substances. In other words, each point on the chromatogram represents the signal intensity of the mass spectrum recorded at that specific time point. As can be seen from Figure 7.39, each chromatogram is a collection of mass spectra.

The *mass spectrum* can be recorded in two ways: either as a profile scan or as a centroid scan. Both ways are used in bioanalysis but have different application areas.

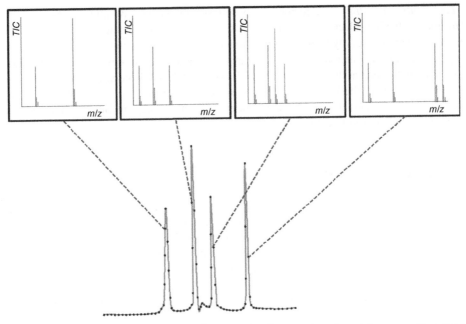

Figure 7.39   *A chromatogram is a collection of points, and each point represents a mass spectrum. In this example, only four mass spectra are shown*

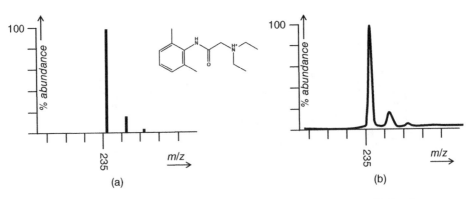

Figure 7.40   *Centroid scan (a) and profile scan (b) of protonated lidocaine*

A *profile scan* shows all the signals that are recorded within a peak. In other words, if 10 scans are performed within a certain mass peak, the peak is shown as a continuous line that connects the scan points. The advantage of using profile scan is the possibility of retrospective qualitative analysis (this will be discussed elsewhere).

A *centroid scan* shows only a bar at the centroid m/z (or center mass) of the mass peak. It collects all the scans within the limits of the mass peak and writes this as one signal with zero peak width. The advantage of recording centroid scans is that it uses much less data space. Figure 7.40 shows both a profile scan and a centroid scan of lidocaine.

Interpretation of mass spectra in LC-MS: There is very little interpretation possible in mass spectra generated from an LC-MS analysis. This is because the soft ionization techniques that are used mainly generate molecular ions. From this, the mass of molecular ion, the elemental composition, the charge of the analyte, and the presence of adducts can be obtained from mass spectra (see Box 7.4).

In general, mass spectra from single analytes are clean spectra. Only a few signals are present. Usually, the m/z value with the highest mass represents the mass of the molecular ion. In some special cases, higher masses than the mass of the molecular ion are seen in the mass spectrum. This is caused by adduct formation. Known adducts are Na^+ ($\Delta m = +23$), K^+($\Delta m = +39$), and NH_4^+($\Delta m = +18$), which are seen in the positive mode, and CH_3COO^-($\Delta m = +59$), which can be seen in the negative mode.

Some information on the elemental composition can be obtained by using the isotope intensities obtained from a spectrum. In general, the number of atoms present in a compound can be calculated using the following equation:

$$n = \frac{I_{m+1}}{A_{m+1} \times I_m} \tag{7.6}$$

where n is the amount of atoms present in the compound, I_m is the intensity of the first isotope, I_{m+1} is the intensity of the second isotope, and A_{m+1} is the relative presence of the second isotope (relative to the first isotope, which is normalized to 1). Both I_m and I_{m+1} are experimentally obtained from the mass spectrum. The value for A_{m+1} is obtained from literature.

The charge (the value of z in m/z) of the ion can be determined from the isotopes. Isotopes ensure that, besides the mass of the mono-isotope, masses occur that are approximately 1, 2, 3, and so on masses higher. The mass increment between the several isotopes is thus approximately 1. This increment can be used to determine the charge of the ion by using the following equation:

$$z = 1/\delta m \tag{7.7}$$

In other words; if the mass spectrum shows an isotope distribution with mass difference (δm) between the isotopes of 0.5, the charge of the ion is 2. When the mass difference between the isotopes is 0.2, the charge is 5.

Box 7.4 Interpretation of a Mass Spectrum Exemplified by Haloperidol

Figure 7.41 shows the mass spectrum of haloperidol together with a description of the most important masses. The molecular ion of haloperidol has an m/z value of 376. This is the mass of the monoisotopic haloperidol + the mass of a hydrogen ($[M + H]^+$). The m/z values 377 and 378 belong to the ^{13}C and ^{37}Cl isotopes, respectively. The intensity of the ^{13}C isotope (I_{m+1}) is 23.5% (compared to 100% signal of the mono-isotopic haloperidol (I_m); both these values are obtained from the spectrum). Because the abundance of the ^{13}C isotope is 0.011 (A_{m+1} – this value is obtained from literature), the amount of C-atoms can be calculated using the equation from above:

$$n = \frac{I_{m+1}}{A_{m+1} \times I_m} = \frac{23.5}{0.011 \times 100} = \text{approximately } 21$$

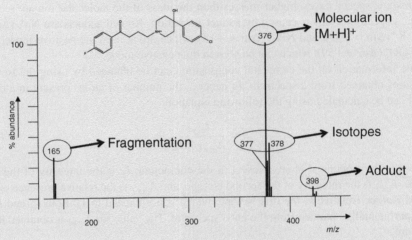

Figure 7.41   *Mass spectrum of haloperidol; its molecular ion (m/z 376), its sodium adduct (m/z 398), as well as ^{13}C (m/z 377) and ^{37}Cl (m/z 378) isotopes*

The same calculation can be performed for chlorine: ($I_{m+1} = 35.4\%, I_m = 100\%, A_{m+1} = 0.3127$).

$$n = \frac{I_{m+1}}{A_{m+1} \times I_m} = \frac{35.4}{0.3127 \times 100} = \text{approximately } 1$$

The m/z value of 398 represents the sodium adduct of haloperidol ($[M + Na]^+ = 375 + 23 = 398$).

The charge of the molecular ion can be calculated from the mass differences between the isotope:

$$z = \frac{1}{\delta m} = \frac{1}{1} = 1$$

There are typically two *measuring modes in single MS* (i.e., when the mass spectrome-ter only records the mass spectrum without fragmentation) used: "full scan" and "SIM". Below, these two scan modes are discussed.

During a *full scan* analysis, the mass spectrometer scans over a range of mass units in a certain time. This is called *scan time* or *scan cycle time*. An example of this would be a scan range between m/z 200 and m/z 500 within 0.5 seconds. This means that every second, two mass spectra are generated. Full scan analysis is typically carried out to obtain qualitative information of the analytes.

When much is known of the analytes to be determined, one can operate the mass spectrometer in the SIM mode. This means that only specific m/z values are monitored. In other words, in case of a SIM analysis, no mass spectra are obtained, only the intensity of a specific mass. Although less spectral information is obtained, SIM is very useful for quantitative purposes. When the mass spectrometer is operated in the SIM mode, much more scan time is related to the specific mass, resulting in a much higher signal-to-noise ratio compared to a full scan analysis. This allows a lower detection limit.

Several *measuring modes in tandem MS* (i.e., when the mass spectrometer is able to fragment the ions that are introduced into the analyzer) can be used: product ion scan, selected reaction monitoring (SRM), multiple reaction monitoring (MRM), precursor ion scan, and constant neutral loss. Mainly, fragmentation can be divided into three steps. In the first step, the m/z value of interest is selected (also referred to as MS1); in the second step, the selected ion is fragmented; and, in the last step, the fragments are detected (also referred to as MS2). Note that it depends on the fragmentation principle of a specific mass spectrometer if one can perform the scan modes as discussed below. Those mass analyzers that perform *fragmentation in space* are able to carry out all the scan modes, whereas those mass analyzers that perform *fragmentation in time* do not allow to carry out precursor ion scan and constant neutral loss.

After fragmentation of the analyte of interest, all fragment ions are determined in a full scan analysis. In other words, all fragment ions that are generated after the fragmentation will be visible in a spectrum. *Product ion scans* are typically run to gain structural information on the analytes. The mass spectrometer is operated such that a selected m/z value is chosen in the MS1 step, whereas, after fragmentation, in the MS2 step a range of m/z values is scanned.

In SRM, only a specific fragment or fragments originating from a selected precursor ion is monitored. Similar to SIM in single MS, only the intensity of one specific or a few specific fragment m/z values are determined. No or little spectral information is obtained. Using SRM, the mass spectrometer is operated in a highly selective manner such that very low detection levels can be obtained. SRM is typically used to quantitate substances that occur in the low picogram level. The mass spectrometer is operated such that in the MS1 step, a selected m/z is chosen, whereas in the MS2 step, the signal of one or several selected m/z values (of the fragments) is determined.

MRM is in principle similar to SRM, but in MRM a specific fragment or fragments originating from several selected precursor ions are monitored. As with SRM also, MRM is used to obtain very low detection limits. The mass spectrometer is operated in the same manner as with the SRM, and the only difference is that in the MRM mode, several m/z values are selected in the MS1 step.

In a *precursor ion scan*, the m/z values of precursor ions are recorded, which produce a selected product ion after fragmentation. In other words, there can be several compounds (with different m/z values) that generate a specific product ion after fragmentation. An example of this is the fragmentation of a drug and its metabolites in an LC-MS/MS analysis. The analytes have a high degree of structural similarity. Although the drug as well as its metabolites have different m/z values, they often generate one or more fragments, which are identical. Precursor ion scan analysis is typically carried out in an exploratory phase to elucidate the metabolism of a drug. The mass spectrometer is operated such that the first analyzer scans a range of *m/z* values, whereas the second analyzer only detects a selected *m/z* value. In this way, only those masses are recorded that produce a selected fragment. Only mass spectrometers that perform *fragmentation in space* can be used to carry out precursor ion scans.

In a *constant neutral loss* scan, all precursor ions are recorded that produce product ions that differ with a constant mass. This mass difference is related to a common neutral fragment, which is not detected in the mass spectrometer. The mass spectrometer is operated such that the first analyzer carries out a scan of all *m/z* values, and the second analyzer (after the ions are fragmented) scans at a specific offset (this offset is the mass of the neutral loss). As with the precursor ion scan, constant neutral loss scan is used to identify closely related compounds in a complex mixture. Only mass spectrometers that perform *fragmentation in space* can be used to carry out constant neutral loss scans.

A *chromatogram* is obtained in all of the above-mentioned measuring modes. This chromatogram not only contains information on the retention time and on concentration by means of the peak height of the analyte, but also contains the mass spectral information. In case of chromatograms obtained from full scan analysis (single MS) and product ion scan, constant neutral loss, and precursor ion scan (all tandem MS), this spectral information contains much data. The spectral information obtained from SIM, SRM, and MRM is minimal. In all cases, the spectral information can be interpreted retrospectively, meaning that data processing can be carried out when the analytical run has finished.

Spectral information from an analyte is interpreted by studying its *mass spectrum*. This mass spectrum is found as a single mass spectrum at the apex of the analyte, or as an average mass spectrum that is obtained by summing mass spectra over a certain area of the chromatographic peak. The interpretation of this mass spectrum is discussed in this chapter.

After an LC-MS analysis of a mixture is carried out, the software allows performing retrospective data extraction. Depending on which type of scan is carried out, the amount of retrospective information varies. An *extracted ion chromatogram* is the chromatogram of a selected *m/z* value or a selected mass range.

For full scan analyses, in either a single mass spectrometer (full scan) or a tandem mass spectrometer (product ion scan), one can simplify a chromatogram with many peaks to a much less complex chromatogram with only a few peaks. These peaks represent those substances with the selected *m/z* value (either the molecular ion in single MS or a selected fragment in tandem MS). The obtained peak height can be used for quantitative purposes. (See Figure 7.42a.)

When high-resolution mass spectrometers are used that are operated in the full scan mode (or product ion scan mode) using profile scans, extraction of ion chromatograms can be carried out in a very selective manner. Because of its high resolution, the *m/z* values used to produce an extracted ion chromatogram can be very narrow. This results in a dramatic decrease of the complexity of a chromatogram (see Figure 7.42b).

A *base peak ion chromatogram* is a particular way of extracting an ion chromatogram retrospectively. The term *base peak* refers to the *m/z* value in a mass spectrum, which has the highest signal intensity. In this way, a *base peak–extracted chromatogram* is the collection of the signal intensities of the *m/z* value with the highest peak. In other words,

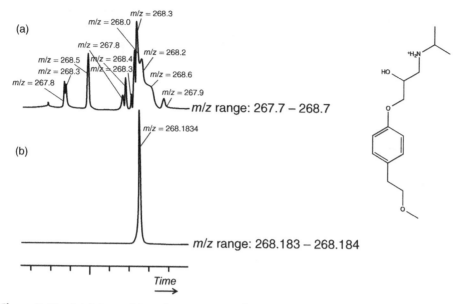

Figure 7.42 *(a) Extracted ion chromatogram of metoprolol obtained with a low-resolution mass spectrometer; and (b) extracted ion chromatogram of metoprolol obtained with a high-*
-resolution mass spectrometer

suppose a 25-minute chromatogram is constructed from 3000 mass spectra (scan speed, 2 scans/second). The signal intensity of the chromatogram at the time point of 2 minutes is obtained from the m/z value with the highest signal intensity in mass spectrum number 240, whereas the signal intensity at the time point of 10 minutes is obtained from the m/z value with the highest signal intensity in mass spectrum number 1200. (See Figure 7.43.) In analysis of complex samples containing many different substances, it is easier to get a visual impression of the chromatographic quality of the analysis using this base peak ion chromatogram compared to the normal TIC (total ion current) chromatogram. In the base peak ion chromatogram, often distinct peaks will be observed, whereas in the TIC chromatogram from the same file, these can be small or even absent.

7.9.7 Pitfalls in LC-MS

The phenomenon of *matrix effects* is one of the largest challenges in LC-MS analysis of drugs in complex biological samples. Due to these effects, the signal intensity of the analyte to be quantified is suppressed or enhanced: either way, the signal is not representative anymore of the concentration of the analyte. This gives bad reproducibility and linearity. Matrix effects are caused by interferences, which elute at the same retention time as the analyte to be determined. The mechanism of matrix effects is still not completely clear and mainly occurs when ESI is used as an ion source. There are two common explanations: (i) there is altered ion desorption from the droplet surface in the electrospray process, and (ii) the analyte competes with the co-eluting interferences for charges. Both processes lead to a change in the signal intensity of the analyte. From an LC-MS chromatogram obtained from

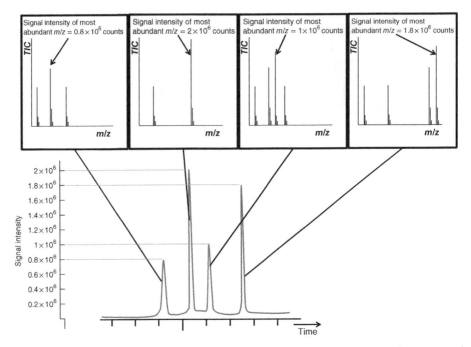

Figure 7.43   *Base peak chromatogram showing four peaks of four different substances with their respective mass spectra and signal intensities*

a complex sample, it will not directly be clear that one is dealing with matrix effects when they occur. This is certainly the case when operating the system in the SIM or SRM/MRM mode. Because in this mode only the ions of interest are monitored, all the other substances that elute from the column (which have different *m/z* values) are not registered. Figure 7.44 illustrates both a full scan trace and a SIM trace of the analysis of a drug in plasma after sample preparation. It can be seen that the SIM analysis yields an uncomplicated chromatogram, whereas the full scan analysis reveals how many compounds elute from the chromatographic column.

It is always recommended to carry out a thorough investigation if matrix effects affect a specific analysis. For this purpose, two approaches can be used: the post-extraction addition method and the post-column infusion method. Both approaches are discussed below.

In *post-extraction addition*, first, a blank biological matrix is prepared according to the sample preparation method used. This yields an extract that does not contain the analyte of interest. To this extract, a certain amount of analyte is added. Second, to a similar volume of extract solvent, the same amount of analyte is added. In this way two samples are produced that contain the same amount of analyte. Both samples are then analyzed on the LC-MS system. (See Figure 7.45.)

In *post-column infusion*, the LC-MS system used to perform the analysis of the analyte of interest is modified with a T-junction and an infusion pump. These are placed after the chromatographic column and before the ESI ion source. During the post-column

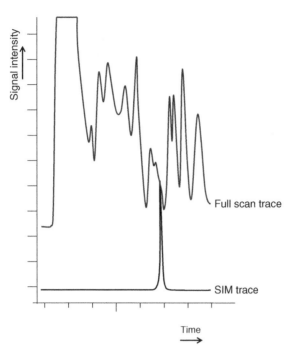

Figure 7.44 *Full scan and selected ion monitoring from same-drug analysis*

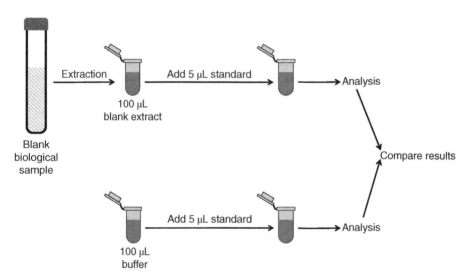

Figure 7.45 *Postextraction addition approach to estimate the effect of the matrix on the analysis*

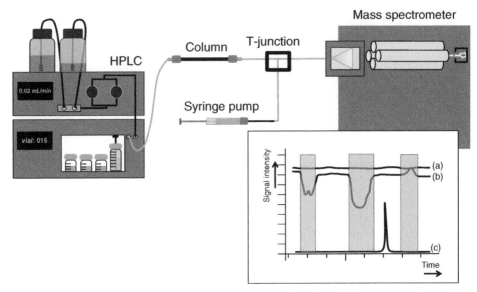

Figure 7.46   *Postcolumn infusion approach to estimate the effect of the matrix on the analysis. The chromatogram shows: (a) the continuous signal of drug X infused by the syringe pump; (b) the continuous signal of drug X infused by the syringe pump after injecting a blank extracted biological sample in the HPLC system; and (c) the signal of drug X injected in the HPLC system. The gray areas indicate the time ranges with signal disturbances*

infusion experiment, the analyte of interest is infused directly from the infusion pump into the post-column LC flow. This generates a continuous signal of the analyte in the mass spectrometer. After sample preparation of a blank biological matrix that does not contain the analyte of interest, the extract is injected into the LC-MS system. At those retention times when matrix effects occur, the constant analyte signal in the MS is disturbed: the signal intensity either goes up (enhancement) or goes down (ion suppression). The chromatogram obtained from this blank analysis with post-column infusion is then compared to a chromatogram of the analyte of interest. When the retention time of the analyte of interest coincides with the areas in the chromatogram where signal disturbance has been observed, the probability of matrix effects is shown (see Figure 7.46).

Because each biological sample is highly individual, it is recommended to carry out the evaluation of matrix effect using at least six separate blank samples.

Cross-talk is a typical challenge in MRM analyses. When scan times are short, fragments of a specific transition might still be in the collision cell at the moment the next precursor ion is being fragmented. In other words, the collision cell is not cleared for all ions at the moment the next precursor ion enters the collision cell. This does not need to be problematic; however, when a similar fragment is monitored in two consecutive MRM scans, the fragment from the first MRM scan might appear in the second MRM scan.

An example of this is the analysis of a certain compound (with a transition $584 \rightarrow 274$) and its d_4 deuterated internal standard (with a transition $589 \rightarrow 274$). Although both transitions are different (because they have different precursor ion masses), the mass of the fragment is the same. The presence of m/z 274 (originating from $584 \rightarrow 274$) in the $589 \rightarrow 274$ scan leads to false signal intensities in this transition. This, in turn, leads to decreased reproducibility and linearity.

Cross-talk can occur when mass spectrometers are used that carry out fragmentation in space. In other words, it can occur in systems like triple quadruple or Q-TOF. Cross-talk will not take place in ion trap analyzers. To check if cross-talk causes challenges in a specific analysis, it is possible to insert a so-called dummy transition during the MRM analysis. This means that a nonexisting/nonsense transition is used where ion 274 is monitored. To use the same example, one could add, for instance, $900 \rightarrow 274$ as a dummy transition after $584 \rightarrow 274$. When a signal is observed in the $900 \rightarrow 274$ transition, one is dealing with crossover.

When using LC-MS to determine the quantity of compounds, often isotopic-labeled internal standards are used. The use of these is advantageous because the isotopic-labeled internal standards exhibit the same physical-chemical properties as the analyte to be determined, except for its m/z. In other words, it is easy to determine the peak intensity of the analyte as well as the peak intensity of the internal standard by using their respective m/z values. In many cases, both analyte- and isotopic-labeled internal standard elute at the same retention time. However, this is not always the case, as, especially in the case of deuterated internal standards, retention shifts can occur. This is called the *deuterium isotope effect on retention time*: when the amount of deuterium atoms increases, the hydrophobicity of

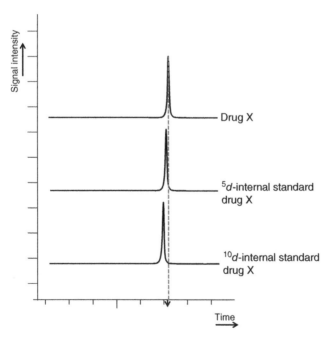

Figure 7.47 *Chromatograms of time-resolved analyte and two deuterated internal standards*

a substance changes slightly. The more deuterium, the shorter the retention time. This is caused by the fact that a $C-^2H$ bond is slightly more polar than the $C-^1H$ bond. In cases when the LC-MS analysis is performed by running segmented analysis (a preprogrammed time-dependent mass analysis), this can become a pitfall.

Figure 7.47 shows the retention times of the ion extracts of an analyte and two different internal standards in a segmented analysis. It is clear that the retention shift of the internal standard into a segment with another *m/z* value causes false low analysis peak intensities of this standard. This will cause erroneous determination of the analyte.

The above-described deuterium effect is less pronounced for ^{13}C-labeled compounds.

8

Gas Chromatography (GC)

Stig Pedersen-Bjergaard

School of Pharmacy, University of Oslo, Norway
School of Pharmaceutical Sciences, University of Copenhagen, Denmark

Gas chromatography (GC) is a technique used for separation and detection of volatile substances or substances that can be converted into volatile derivatives. In GC, a microliter volume of sample is injected into a heated injection port, where the sample constituents immediately are evaporated. The sample vapor is then transferred with the carrier gas from the injection port and into the separation column (GC column). The inner wall of the GC column is coated with a thin layer of stationary phase. The different sample constituents are transported with the carrier gas through the GC column. However, their velocities are different, depending on their volatility and their degree of interaction with the stationary phase. Compounds with low volatility and strong interaction with the stationary phase will be strongly retained in the GC column, and move slowly along the column. Thus, different sample constituents are physically separated based on different volatility and different affinity toward the stationary phase. At the end of the GC column, a detector is located to detect and measure the individual sample constituents.

8.1 Basic Principles of GC

The basic principle of GC is illustrated in Figure 8.1. A microliter volume of sample is injected by a small syringe into the *injection port* on top of the GC. The injection port is heated, and the sample constituents immediately evaporate inside the injection port.

The evaporated sample constituents are then transported along the *capillary column* with the *carrier gas*. The carrier gas is the mobile phase in GC. The capillary column (=*GC column*) is a long and open tube with a typical internal diameter of 0.25 mm. The inside wall of the GC column is coated with a thin layer of stationary phase. The sample constituents are interacting with the stationary phase as they pass through the GC column, and different

Bioanalysis of Pharmaceuticals: Sample Preparation, Separation Techniques and Mass Spectrometry,
First Edition. Steen Honoré Hansen and Stig Pedersen-Bjergaard.
© 2015 John Wiley & Sons, Ltd. Published 2015 by John Wiley & Sons, Ltd.

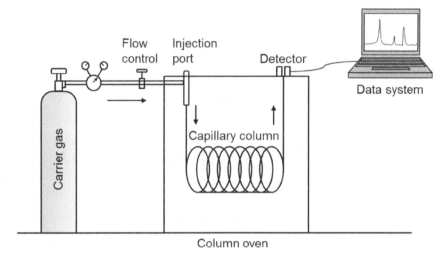

Figure 8.1   *Schematic illustration of a gas chromatograph*

compounds are separated based on differences in their affinity toward the stationary phase. At the end of the GC column, the separated compounds are measured by a detector. The detector signal is measured continuously during the separation, and this signal is recorded as a chromatogram by a computer. The entire GC column is located inside an oven (GC oven), because temperature is a very important parameter controlling the retention in GC.

In GC, analytes must be volatile or semivolatile in order to be transported with the carrier gas to the detector. Consequently, only volatile and semivolatile substances can be separated and detected by GC. The volatility of a substance is temperature dependent, and the vapor pressure of the substance increases with temperature. Therefore, retention is reduced in GC by increasing the GC oven temperature. Highly volatile substances are separated at low temperatures, whereas less volatile compounds have to be separated at higher temperatures. Due to the volatility criterion, bioanalytical GC is limited to small-molecule drug substances with molecular weights below approximately 500. Two additional issues further limit the applicability of GC, namely, the analyte polarity and the thermal stability. Polar compounds are difficult to separate by GC because they are prone to serious peak tailing. Also, the thermal stability of the analyte plays a role. Because GC is accomplished at elevated temperatures, the analytes have to withstand temperatures up to 250–300 °C in order to be GC amenable. Consequently, nonpolar small-molecule drug substances are ideal for GC, whereas more polar substances and large-molecule substances are ideal for *liquid chromatography* (LC). Most pharmaceuticals and biopharmaceuticals are indeed either relatively polar or large-molecule substances, and therefore LC is much more in use in the bioanalytical laboratory than GC.

8.2   GC Instrumentation

A photo of a GC is shown in Figure 8.2, and a schematic illustration is shown in Figure 8.1. The GC is connected to a computer and fully controlled by software.

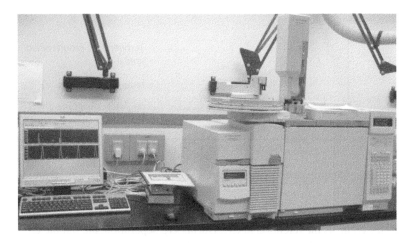

Figure 8.2 *Photograph of a gas chromatograph*

The carrier gas used for GC is normally contained in a high-pressure steel cylinder, and it is delivered through a reduction valve and into the GC instrument. The carrier gas is heated and passes through flow controllers into the injection port. The heated carrier gas flows through the injection port and the GC column, and to the detector.

The *reduction valve* brings the carrier gas pressure from the cylinder down from 200 to about 5 bars as the operating column inlet pressure. The *flow controllers* are used to set the flow rate of carrier gas and to make sure that this flow remains constant over time. Operation at constant flow of carrier gas is mandatory in order to perform repeatable GC separations with repeatable retention times and detector signals. The flow rate of carrier gas is controlled from the computer software. In bioanalysis, the samples are usually dissolved in a volatile organic solvent prior to GC analysis. Often, this solvent is the extraction solvent from liquid–liquid extraction (LLE) of a biological fluid. The samples are injected with a microsyringe into the GC. Common injection volumes are 0.5–2.0 µl. The temperature of the injection port is kept high enough to evaporate the sample constituents immediately, and the carrier gas transfers the evaporated sample constituents into the GC column. Typically, injection port temperatures are in the range of 200–300 °C. The injection port temperature is normally higher than the initial GC column temperature to obtain a rapid transfer of sample constituents into the column. If the injection port temperature is too low, slow and incomplete transfer of sample constituents may occur. In contrast, if the injection port temperature is too high, this may result in decomposition of sample constituents during injection. In other words, the injection port temperature has to be balanced and optimized for each method. The injection port temperature is controlled from the computer software.

A detector is placed at the outlet of the GC column, and the detector is heated typically to 250–350 °C. This is generally 25–50 °C higher than the final (and highest) GC column temperature used for separation. The detector is heated in order to avoid condensation of sample constituents in the column end, just prior to detection. Also, heating the detector prevents sample constituents of low volatility from condensing and contaminating the inside of the detector. The detector temperature is also controlled from the computer software.

The GC column is placed in the GC oven with a heating element and a fan that ensures efficient circulation and temperature control during the analysis. The temperature in the

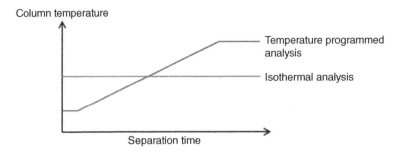

Column temperature

Temperature programmed
analysis

Isothermal analysis

Separation time

Figure 8.3 *Illustration of an isothermal and temperature programmed gas chromatography*

oven is controlled from the computer software. The temperature of the column is the main
parameter that controls the retention in GC. During GC separation, the sample constituents
are distributed between the carrier gas and the stationary phase, but they only move forward through the GC column when they are in the carrier gas phase. The more sample
constituents are distributed into the stationary phase, the longer they are retained in the GC
column. An increase in the column temperature will increase the volatility of the substances
and in this way change the distribution in favor of the carrier gas phase. Therefore, reducing retention in GC is obtained by increasing column temperature. As a rule of thumb, the
retention time is halved when the column temperature is increased about 30 °C. In general,
sample constituents are separated at substantially lower column temperatures than their
respective boiling points.

The GC oven can be operated in two different ways during the GC separation: in *isothermal mode* (*isothermal analysis*) and in *temperature programmed mode* (*temperature programmed analysis*). In isothermal mode, the temperature in the GC oven (and in the GC
column) is kept constant during the entire separation, as illustrated in Figure 8.3. In temperature programmed mode, the temperature in the GC column is gradually increased during
the separation (see Figure 8.3).

Samples of low complexity and with constituents of similar volatility can be separated
using the isothermal mode. However, in the bioanalytical laboratory, samples are often
more complicated, containing a broad variety of compounds of different volatility, and
in those cases temperature programmed analysis is preferred. In temperature programmed
analysis, the column temperature is low and typically in the range of 50–100 °C during sample injection. At this stage, only the most volatile sample constituents are carrier through the
GC column. Less volatile sample constituents are condensed at the inlet of the GC column.
After a couple of minutes, the GC oven temperature is gradually increased, typically at a
rate of 5–20 °C/min, and gradually less and less volatile sample constituents are released
from the GC column inlet and carried to the detector by the carrier gas. Thus, even for samples containing compounds with major differences in terms of volatility, very efficient GC
separations can be obtained, as illustrated in Figure 8.4. In the upper chromatogram, separation was accomplished by isothermal GC at 150 °C. The first peaks are narrow, but they
are not very well separated. The less volatile compounds elute as very broad peaks at extensively long retention times. With temperature programmed analysis, in contrast, both the
volatile and less volatile compounds were separated with high efficiency. During temperature programmed analysis, the program is always finished at high temperature (250–325 °C)
to make sure that sample constituents of low volatility are removed from the column. After

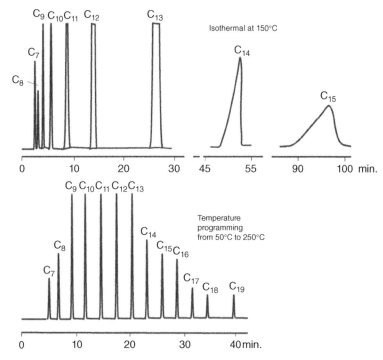

Figure 8.4   *Isothermal versus temperature programming in gas chromatography analysis of n-alkanes*

finishing the temperature program, the GC oven has to be cooled to the initial starting temperature again before a new sample can be injected. Isothermal GC is similar to isocratic elution in LC, whereas temperature programmed GC is similar to gradient elution in LC.

8.3   Carrier Gas

The carrier gas is an inert transport medium for the sample constituents in GC. Thus, in contrast to LC, where the chemical composition of the mobile phase plays an extremely important role for the separations, GC separations are more or less the same with different carrier gases. Carrier gasses like *nitrogen*, *helium*, and *hydrogen* are most frequently used. Short analysis times are desirable, and this is achieved by using a high carrier gas velocity (flow rate) through the GC column. However, the chromatographic separation efficiency is dependent on the carrier gas velocity, and the optimum gas velocity is different for nitrogen, helium, and hydrogen. This is shown in Figure 8.5, where the height equivalent to a theoretical plate (HETP) is plotted against the mean gas velocity (cm/s) through the column in a *van Deemter plot* (also discussed in Chapter 4). Remember that the separation efficiency of the system increases with decreasing HETP value.

According to Figure 8.5, the optimal linear velocity for nitrogen is approximately 20 cm/s, whereas for hydrogen the corresponding value is about 40 cm/s. Consequently, using optimal carrier gas velocities, the analysis time is halved by using hydrogen as carrier gas compared to the use of nitrogen. When using flow rates higher than the

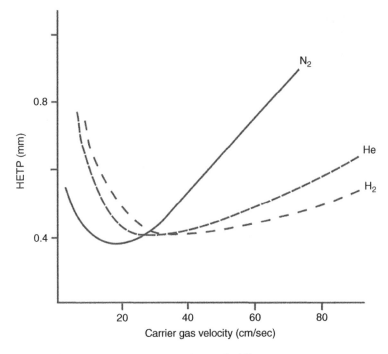

Figure 8.5 *van Deemter plots with different carrier gases*

optimum, it is important to note that the steepnesses of the van Deemter plots are different for the different carrier gases. For nitrogen, increasing the flow rate above the optimum will result in a significant loss of separation efficiency. For hydrogen, the loss in separation efficiency is less. The steepness expresses the readiness in mass transfer between the mobile and the stationary phases. The nitrogen molecule is relatively large, and analyte molecules are more likely to bump into nitrogen molecules than into the small hydrogen molecules. This implies that nitrogen will hamper mass transfer and this is resulting in broader peaks. Helium is safer to use than hydrogen and is therefore preferred as a carrier gas. The analysis time is reduced by 40% compared to nitrogen.

Hydrogen and helium are the most popular carrier gases. The optimal linear carrier gas velocities in the range of 20–40 cm/s are equivalent to volumetric flow rates of approximately 1–2 ml/min with narrow capillary columns. The carrier gases used are of high purity, with a low content of oxygen and organic impurities. This is extremely important in order to avoid deterioration of the stationary phase (caused by oxygen) and to avoid high background in the detector response (caused by organic impurities). The purities of GC carrier gases in use normally exceed 99.99%. Although the consumption of carrier gas is relatively low, the prices for the gases are high due to the high purity requirements.

8.4   Stationary Phases

The stationary phase in GC is coated as a thin film on the inside wall of the capillary column. The stationary phase is either a nonvolatile liquid or a solid material. A solid stationary

phase is an organic or an inorganic polymer, and separation is obtained either due to different adsorption to the surface of the polymer or by sieving through well-defined pores of the
polymer. GC based on solid stationary phases is named *gas–solid chromatography* (GSC).
GSC is primarily used to separate gases and very volatile substances, and it is little in use in
the bioanalytical laboratory. In bioanalytical GC, the stationary phases are usually based on
nonvolatile organic liquids, and this is termed *gas–liquid chromatography* (GLC). In GLC,
separation is based on the distribution of sample constituents between the carrier gas and
the liquid stationary phase. Liquid stationary phases in GC are temperature-stable liquids
that have very low vapor pressure. Many different stationary phases have been developed
throughout the years, but the preferred stationary phases in modern GC are either *polysiloxanes* or *polyethylene glycols*. These are liquids at room temperature, chemically stable up to
225–325 °C, and with low vapor pressure at high temperatures. The stationary phase should
be a liquid throughout the entire operational temperature range. Chemical stability at high
temperatures is important because most GC is accomplished in temperature programmed
mode, and here the GC analysis often finishes at 250–325 °C. Low vapor pressure is important in order to avoid excessive *column bleeding* of stationary phase from the column during
the high-temperature operation.

One example of a polysiloxane stationary phase is illustrated to the left in Figure 8.6,
where the skeleton of the polydimethylsiloxane polymer is shown. Polydimethylsiloxane
is the most hydrophobic of the polysiloxanes, and this stationary phase is frequently used
in general columns for separation of nonpolar sample constituents.

The polarity of the GC stationary phase can be varied by substitution of some
of the methyl groups with either phenyl or cyanopropyl groups, as illustrated in
Figure 8.6. This provides medium-polar columns termed *polyphenylmethylsiloxane* or
polycyano-propylmethylsiloxane. The level of substitution can vary significantly, and a
large number of different stationary polysiloxane phases exist. With increasing degrees of
phenyl substitution, the polarity of the stationary phase will increase. When cyanopropyl
groups are introduced, the polarity is increased even further.

Polysiloxanes are very temperature stable and can be used over a wide temperature range.
However, the introduction of polar cyanopropyl groups reduces the temperature stability
of the stationary phase to some extent. Thus, whereas polydimethylsiloxane columns can
be used up to 300–325 °C, polycyanopropylmethylsiloxane columns can only be used up
to about 275 °C.

Polyethylene glycol represents another family of stationary phases. The general skeleton
of polyethylene glycol polymers is shown in Figure 8.7.

Polyethylene glycols are also named Macrogol or Carbowax followed by a number indi-
cating the molecular weight. The temperature stability increases, and the vapor pressure

Figure 8.6   *The basic skeleton of polydimethylsiloxane, polyphenylmethylsiloxane, and polycyanopropylmethylsiloxane. The numbers of repeating units (n, p, and x) can vary significantly
form one stationary phase to another*

$$HO-CH_2-CH_2 \left[O-CH_2-CH_2 \right]_m OH$$

Figure 8.7   *The basic skeleton of polyethylene glycol*

decreases with increasing molecular weight. These polymers are available with different molecular weights, but the most popular polymer for GC is Carbowax 20M, which has a molecular weight of 20 000. It can be used in the temperature range from 60 to 250 °C. The hydroxyl groups terminating the polyethylene glycol chain are polar, and these are responsible for the polar characteristics of the stationary phase. Polar GC stationary phases are used for separation of polar sample constituents. They also provide different separation selectivity as compared to polysiloxane-based stationary phases. This will be discussed in detail in Section 8.5.

GC columns all have an operational temperature range. The upper temperature limit is important not to exceed in order to avoid column bleeding (loss of stationary phase) and degradation of the stationary phase. Column bleeding and degradation of the stationary phase will seriously compromise the separation efficiency of the column. GC columns also have a lower temperature limit. Below the lower temperature limit, the stationary phase becomes too viscous and the diffusion of the analyte will be reduced, leading to broader peaks and poor chromatography.

8.5   Separation Selectivity in GC

As mentioned, the retention of substances is governed by the temperature of the column. Increasing the column temperature increases the volatility of the substances, and the retention decreases. Interaction between the analyte and the stationary phase is another important parameter influencing the retention. Strong intermolecular interactions result in increased retention. Heptane and water have similar boiling points, 98 and 100 °C, respectively. If they are to be separated in GC on a nonpolar stationary phase like polydimethylsiloxane, the water molecules will pass through the column with the carrier gas without significant retention. The reason for this is that water molecules preferably interact through hydrogen bonding interactions, but polydimethylsiloxane cannot participate in these interactions. In contrast, the heptane molecules will interact with polydimethylsiloxane through van der Walls interactions (hydrophobic interactions) and will be retained by the stationary phase. If the stationary phase is changed to a more polar one based on polyethylene glycol, the water molecules can now interact through hydrogen bonding with the stationary phase and will be retained in the column. In this case, the heptane molecules will have less retention.

A very nonpolar stationary phase (like polydimethylsiloxane) principally interacts with sample constituents through hydrophobic interactions. Hydrophobic interactions are relatively weak, and therefore the nonpolar stationary phases provide relatively weak retention. The interactions become stronger as the polarity of the stationary phase is increased, because polar interactions are involved. The polar interactions can be dipole–dipole interactions or hydrogen bonding interactions. Dipole–dipole interactions are most important when using medium-polar polysiloxanes as the stationary phase, whereas hydrogen bonding interactions are more dominant when using polar polyethylene glycol stationary phases.

When using the very nonpolar stationary phases (like polydimethylsiloxane), neither dipole–dipole interactions nor hydrogen bonding interactions are in force. Therefore, on nonpolar GC columns, the retention of a given solute is a function of its volatility and its ability to interact with the stationary phase based on hydrophobic interactions. Introduction of phenyl groups in the stationary phase increases the retention of substances due to the increase in polarity. The phenyl groups give rise to intermolecular interactions due to temporary dipoles. Retention increases further with increase in a percentage of phenyl groups. Introduction of cyanopropyl groups into the stationary phase also increases retention. In the latter case, dipole–dipole interactions can occur between solutes and the stationary phase. Thus, on medium-polar GC columns, the retention of a given solute is a complex function of its volatility and its ability to interact with the stationary phase based on temporary dipole, dipole–dipole, and hydrophobic interactions. Polyethylene glycol also provides hydrogen bonding interactions with analytes, especially compounds having an alcohol functional group, and alcohols will have a relatively stronger retention compared to other analytes. Thus, on polar GC columns, the retention of a given solute is a complex function of its volatility and its ability to interact with the stationary phase based on hydrogen bonding, dipole–dipole, and hydrophobic interactions.

As discussed, retention depends on both the boiling point of the substances and the interactions with the stationary phase. The consequence is that different stationary phases provide different selectivity, and the order of retention of analytes can thus be changed by changing the stationary phase. An example of this is shown in Box 8.1, where three different solutes with very similar boiling points have been separated on three very different GC columns.

Box 8.1 Retention Times of Benzene, Cyclohexane, and Ethanol on Three Different Stationary Phases

Benzene (B_p = 80 °C) Cyclohexane (B_p = 81 °C) Ethanol (B_p = 79 °C)

	Benzene	Cyclohexane	Ethanol
Polydimethylsiloxane	8.00	8.32	2.68
Polycyanopropyl-phenyl-dimethylsiloxane	8.69	8.10	3.47
Polyethylene glycol	6.46	2.27	6.46

On the nonpolar polydimethylsiloxane, the retention of ethanol is low. Ethanol is very polar, and the tendency for hydrophobic interactions is relatively low. In contrast, the hydrocarbons benzene and cyclohexane are more strongly *retained by hydrophobic interactions due to low polarity. On the medium-polar 6% cyanopropyl-phenyl 94% dimethylpolysiloxane, the retention of ethanol has increased due to dipole–dipole interactions between the $-C\equiv N$ and $-OH$ groups. The medium-polar stationary phase contains 94% dimethyl, and therefore benzene and cyclohexane are still retained by hydrophobic interactions. On the medium-polar stationary phase, the retention order of benzene and cyclohexane has changed as compared to the nonpolar stationary phase. Benzene is now more strongly retained due to π–π interactions between benzene itself and the benzene rings in the stationary phase. With the polar column containing polyethylene glycol, the retention of ethanol is strong. This is principally due to hydrogen bonding interactions between the $-OH$ groups of ethanol and the $-OH$ groups of polyethylene glycol. The retention of cyclohexane is now very low, whereas benzene with π-electrons is still retained strongly by the polar stationary phase.*

8.6   Columns

In GC, two different types of columns are used:

• Capillary columns
• Packed columns.

Originally GC instruments used packed columns, but today bioanalytical GC separations are performed using capillary GC instruments.

Capillary columns can be made of a number of materials like metal or fused silica, but fused silica capillary is preferred due to its robustness and inactivity. Fused silica is very fragile, and therefore the outer surface of the fused silica capillary is coated with a polyimide layer to improve the mechanical stability. This results in very flexible columns that are easy to handle in daily practice. The inner diameter of the columns is in the range of 20–500 µm, but most often columns with 250 or 320 µm inner diameter are used. The stationary phase is coated as a thin film on the inside of the column. The stationary phase can either be physically coated on the inner surface or be chemically bonded to the surface. The more stationary phase coated in the column, the greater the film thickness will be, and the greater the retention will be. The film thickness is in the range from 0.05 to 10 µm, and the typical film thickness is 0.25 µm. Columns with thin film are often used in the analysis of high-boiling substances as thin film reduces retention and provides acceptable analysis times, without using too-high column temperatures. Thick film columns provide higher retention, and are often used for separation of more volatile substances.

The capillary GC columns are open tubes, and there is only little pressure drop across the column when carrier gas flows through. Therefore, it is possible to use fairly long capillary columns. In bioanalytical applications, the typical column length is in the range of 10–30 m. As a rule of thumb, the number of theoretical plates (N) on a capillary GC column is about 3000 per m. A column length of 30 m will therefore provide about 90 000 theoretical plates. This makes GC on capillary columns one of the most powerful separation techniques, and one can detect 50–100 peaks or even more in a single chromatogram.

To fully describe a GC column, both the length and the internal diameter of the column should be given, and also the type of stationary phase and the film thickness should be specified. The length of the column affects the separation efficiency. Longer columns provide a higher number of theoretical plates and better separation efficiency. On the other hand, long columns result in relatively long separation times. The separation efficiency is also affected by the inner diameter of the column. The smaller the inner diameter, the higher the separation efficiency. In contrast, narrow columns are more easily overloaded with sample, and the sample capacity is limited.

8.7   Injection Systems

In GC, it is common to inject the sample constituents dissolved in a volatile solvent. The solvent and the sample constituents evaporate in the injector, and the gas mixture is brought to the column by the carrier gas. Samples are injected with a micro-syringe, and regular injection volumes are in the range of 0.5–2.0 µl. It is important that the sample constituents evaporate instantly and that the sample volume brought to the column does not overload the column. Otherwise, peak broadening will occur.

Injection of 1 µl sample solution in the injector will after evaporation form about 0.5–1.0 ml of gas. The injector must therefore have a sufficiently large volume to allow the sample solution to expand to this gas volume. Moreover, the injection system must prevent the formed gas from entering directly into the capillary column. The internal volume of the capillary is so small that 1 ml of gas will fill up large parts of the column and cause overloading.

In order to prevent overloading, two different injection systems are available:

- Split injection
- Splitless injection.

Split injection and *splitless injection* use the same injector design. Figure 8.8 shows an illustration of a split/splitless injector. The sample solution is injected with a micro-syringe into the injector. The syringe needle penetrates a silicon membrane (*septum*), and the tip of the syringe needle is inserted into a glass tube (*liner*) in the injector. Sample solution is injected and evaporates in the liner. The inner volume of the liner is large enough to allow expansion of the sample into gas phase. The heated carrier gas enters the injector, which has three outlets. The GC capillary column is attached to one outlet. Another outlet, called the *septum purge outlet*, is located just under the septum to prevent any impurity from the septum from reaching the column. The third outlet is called the *split purge outlet*, and this outlet is used during split injection to divert part of the sample volume to waste. Gas flow rates through the three outlets are controlled, and these flow rates determine the amount of sample introduced into the column.

The principle of split injection is that the evaporated sample is split, and only a small portion of the sample is transferred into the column. The rest of the sample is diverted to waste through the split purge outlet. The flow rate ratio between carrier gas to the column and carrier gas through the split purge outlet is called the *split ratio*. With a flow of carrier gas of 1 ml/min through the column, and 50 ml/min through the split purge outlet, the split ratio is 1:50, and only about 2% of the sample enters the column. In this case, 98% is vented

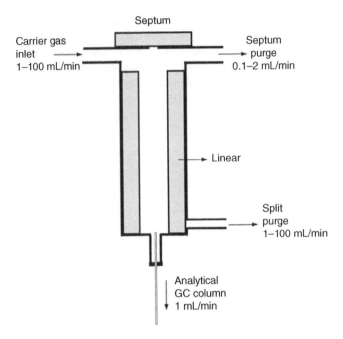

Figure 8.8   *Split or splitless injector for a capillary gas chromatograph*

through the split purge outlet. Split ratios are normally in the range from 10:1 to 100:1. Because only a small portion of the sample volume is used for the separation and detection, split injection is best suited for the analysis of sample solutions containing relatively high concentrations of analytes.

In splitless injection, the total sample volume is introduced into the column. The principle is that the sample evaporates in the injector, and the sample constituents are transferred with the carrier gas to the top of the column where the sample constituents are condensed and concentrated. In splitless mode, the split purge outlet is therefore closed during injection. If 1 μl sample expands to 1 ml of gas and the carrier gas velocity through the column is 1 ml/min, it takes approximately 60 seconds to transfer the sample to the column. To prevent spreading of the analytes throughout the capillary column during the 60 seconds of sample transfer, the analytes must be concentrated at the column entrance. This is accomplished by the *solvent effect*, which is obtained when the solvent used for dissolving the sample constituents condenses in the first part of the column, forming a thin film. The analytes are trapped and dissolved in the condensed solvent. The temperature of the column has to be about 20–50 °C below the boiling point of the solvent to obtain the solvent effect. After the analytes have been trapped in the solvent, the column temperature is gradually increased by temperature programming, and the solvent as well as the sample constituents will gradually elute and be separated based on volatility and interactions with the stationary phase. With this technique, low limits of detection can be obtained because the full injection volume is used for detection.

8.8   Detectors

There are a large number of detectors available in GC, but only the standard detectors used in bioanalysis will be described in detail here. In this section, the flame ionization detector (FID), the nitrogen–phosphorus detector (NPD), and the electron capture detector (ECD) will be discussed. In Section 8.10, the use of mass spectrometry as a detector in GC will be discussed.

8.8.1   The Flame Ionization Detector (FID)

In the FID, which is located at the outlet of the capillary column, the analytes enter a flame and are ionized, and the ions are detected as an electrical current between two electrodes in the detector house. Figure 8.9 shows an illustration of the detector.

The carrier gas leaving the column is mixed with hydrogen at the entrance of the detector, while air is led into the detector house. The carrier gas–hydrogen mixture containing the analytes enters the detector house through a jet. The flame is burning on the top of the jet, and when the organic analytes burn in the flame, ions and free electrons are formed. The amount of charged particles is measured by applying a voltage of 300 V between the jet and the collecting electrode just above the flame. The current measured is proportional to the amount of organic matter that burns.

The FID is the most commonly used GC detector. It gives a linear response for organic compounds over a wide concentration range. The lower limit of detection is about 10^{-9} g.

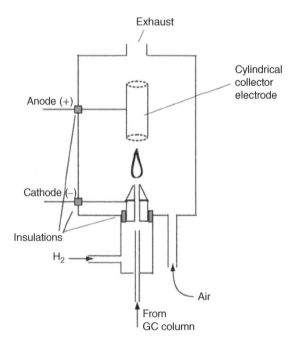

Figure 8.9   *Schematic illustration of a flame ionization detector*

The FID is a universal detector and responds to all organic compounds. Often, more selective detectors are preferred in bioanalysis, and therefore the FID is not in frequent use for bioanalytical applications.

8.8.2   The Nitrogen–Phosphorus Detector (NPD)

The NPD provides a selective response for organic substances containing nitrogen or phosphorus. The detector is also called *alkali flame ionization detector* (AFID) or *thermionic detector* (TID). The detector is built up similar to the FID, but with a crystal of alkali metal salt placed just above the flame (Figure 8.10). The alkali metal salt is usually made of a rubidium salt. Organic compounds containing nitrogen or phosphorus react with alkali metals during combustion under the formation of anions such as cyanide- and phosphorus-containing anions as well as electrons. The response of nitrogen-containing organic compounds can be up to 10^6 times higher than the response of organic substances without nitrogen. The lower limit of detection for nitrogen- and phosphorus-containing organic substances is about 10^{-10} g.

The NPD is more popular in bioanalysis than the FID because it is a selective detector and because many drug substances contain nitrogen.

8.8.3   The Electron Capture Detector (ECD)

The *electron capture detector* (ECD) (Figure 8.11) is a selective detector suited for the determination of compounds with a high affinity for electrons. The detector house contains

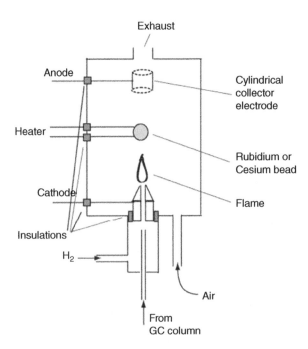

Figure 8.10   *Schematic illustration of a nitrogen–phosphorous selective detector*

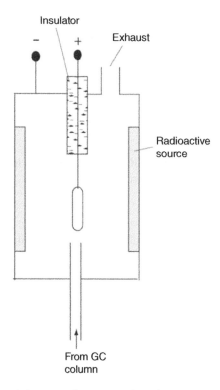

Insulator

Exhaust

Radioactive
source

From GC
column

Figure 8.11 *Schematic illustration of an electron capture detector*

a radioactive foil typically with ^{63}Ni emitting β-electrons that, by collision with the carrier gas molecules, generate a plasma of electrons. Thus, a constant current in the detector provides a background signal that is changed when electron-capturing compounds enter the detector. Compounds containing halogens or nitro groups have high affinities for electrons, and the lower limit of detection is at the femtogram level. The detector is applied in pharmaceutical analysis and bioanalysis where analytes have been derivatized with polyhalogenated reagents. If the analytes do not contain halogens, a number of derivatizing reagents containing fluor are available to make analytes' electron capture active.

8.9   Derivatization

Many drug substances contain polar functional groups such as hydroxyl groups, phenolic groups, amino groups, or carboxylic acid groups. These groups will lower the volatility of the substances due to intermolecular forces like hydrogen bonds or ionic interactions. Furthermore, the polar substances often interact with surfaces in the GC system by adsorption, resulting in poor chromatographic properties and peak tailing. Problems with low volatility and poor chromatographic properties can be suppressed or eliminated by derivatization. The purpose of derivatization is to create more volatile and thermally stable derivatives of

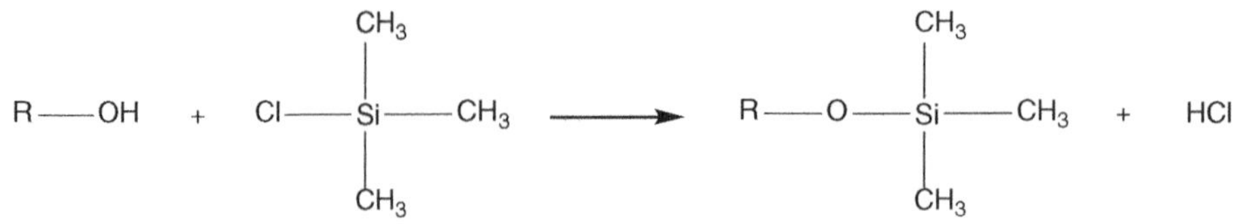

Figure 8.12   *Silylation of a hydroxyl group*

polar substances and to improve their GC behavior and detectability. The derivatization blocks polar intra- and intermolecular forces.

8.9.1   Silylation

By *silylation*, active H-atoms are replaced by trimethylsilyl groups, as shown in Figure 8.12.

The chemical groups such as −OH, −COOH, −NH$_2$, and −NH− are well suited for silylation. Derivatives are thermally stable, volatile, and suitable for GC. A large number of silylating reagents are available from different suppliers.

8.9.2   Acylation

Acylation replaces an active hydrogen atom with an acyl group. Both anhydrides and acid chlorides are used as reagents, and groups such as −OH, −NH$_2$, and −NH can be derivatized. The derivatization reagents usually contain fluorine atoms because fluorine atoms provide stable and volatile derivatives suitable for GC. Commonly used reagents are trifluoroacetic acid anhydride or anhydrides of pentafluoropropionic acid or heptafluorobutyric acid. Amino acids are nonvolatile due to the strong intermolecular ionic interactions. In Figure 8.13, the derivatization of an amino acid is shown involving acylation and alkylation.

8.10   Gas Chromatography–Mass Spectrometry (GC-MS)

8.10.1   Introduction to GC-MS

In *gas chromatography–mass spectrometry* (GC-MS), a mass spectrometer is used as a detector. This is very similar to *liquid chromatography–mass spectrometry* (LC-MS), discussed earlier. However, whereas ionization is performed at atmospheric pressure outside the mass spectrometer in LC-MS, the ionization in GC-MS takes place inside the mass spectrometer under vacuum conditions. Thus, after GC separation, the analytes are bombarded with electrons (electron ionization) in vacuum, and initially this leads to some of the molecules being ionized to either positive ions (molecules lose an electron) or negative ions (molecules take up an electron). These ions are termed *molecular ions*. The mass of a molecular ion is equivalent to the mass of the original molecule because the mass of an electron that is either lost or taken up is incredibly small. Molecular ions are unstable, and they very quickly split into smaller fragments as chemical bonds in the molecules are

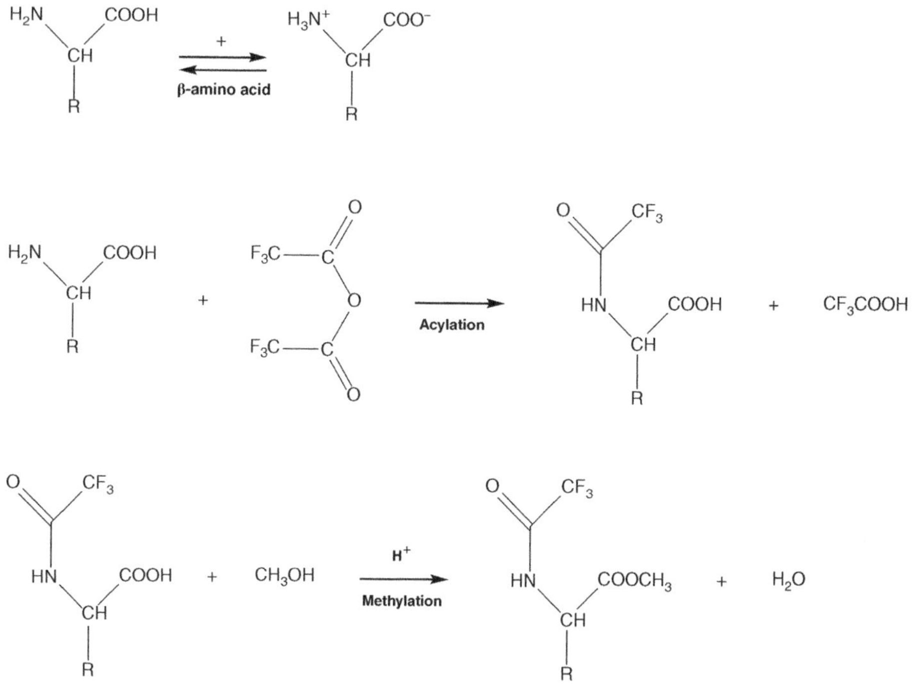

Figure 8.13   *Derivatization of an α-amino acid with trifluoroacetic acid anhydride and methanol*

broken. This process is called *fragmentation*. Some of the fragments will remain ionized (*fragment ions*), whereas others lose their charge. Figure 8.14 shows an example of how the ionization and fragmentation can take place. When chlorambucil (antineoplastic drug) is bombarded with electrons inside the mass spectrometer, some of the molecules ionize to positively charged molecular ions with mass 303. A large proportion of these will be unstable and will fragment. Fragmentation of chlorambucil occurs due to several reactions, but only one of them will be discussed at this stage. The dominating process is the cleavage of CH_2Cl from the molecular ion. The mass of CH_2Cl is 49, and loss of this fragment from the molecular ion gives fragment ions with mass $303 - 49 = 254$.

Ionization and the subsequent fragmentation take place in the *ion source* inside the vacuum region of the mass spectrometer. As mentioned this is in contrast to LC-MS, where ionization principally occurs at atmospheric pressure outside the mass spectrometer. The mass of the molecular ions and the molecular fragments that are still charged (fragment ions) are then determined as the ions are accelerated out of the ion source and into the *mass analyzer*. In the mass analyzer, ions are separated according to the ratio between the mass (m) and the charge (z) in an electrostatic field and are subsequently measured with the detector, as discussed in Chapter 7.

Figure 8.15 shows the electron ionization mass spectrum of chlorambucil (chlorambucil was featured in Figure 8.14). A peak is observed at mass 303 from the molecular ion (M^+) corresponding to the molecular mass of the compound. In addition, the large peak of mass

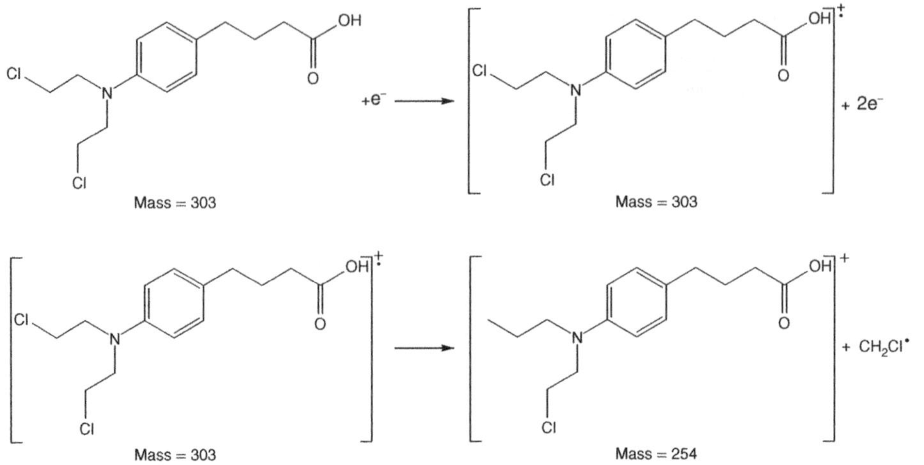

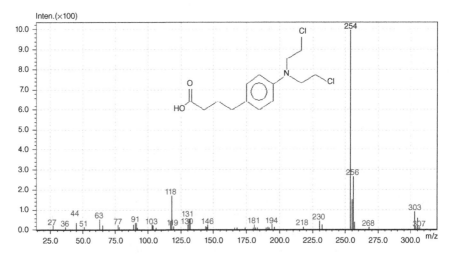

Figure 8.14   *Electron ionization (upper reaction) and one major fragmentation (lower reaction) during gas chromatography–mass spectrometry (GC-MS) of chlorambucil*

Figure 8.15   *EI mass spectrum of chlorambucil*

254 corresponds to the loss of CH_2Cl (M-49)$^+$, and the small peak at mass 268 corresponds to the loss of Cl (M-35)$^+$.

What type of information can we get from an electron ionization (EI) mass spectrum? EI mass spectra can be used to identify organic compounds because different substances have different EI mass spectra. An unknown compound can thus be identified if the EI mass spectrum of the compound exactly matches the EI mass spectrum of a known chemical reference substance. If this is not the case, the compound cannot be identified unambiguously, but GC-MS can still provide important structural information because interpretation of EI mass spectra can provide information about the following:

- Molecular mass
- Elemental composition
- Empirical formula
- Functional groups.

This will be discussed in more detail in this chapter. In addition, GC-MS can be used for quantitative analysis; this will also be discussed at the end of this chapter.

8.10.2   GC-MS with Electron Ionization (EI)

GC-MS is normally performed using EI. During EI, the substances eluting from the GC are bombarded with electrons under vacuum conditions inside the mass spectrometer. The electrons are emitted from a small wire (*filament*) of rhenium or tungsten. The electrons released from the filament are accelerated in an electric field with a potential of 70 V, which means that the electrons gain the energy of 70 eV. When the analytes enter this electron beam, some of the molecules will lose an electron as a consequence of electrical repulsion (negative charges on the electrons are repelling each other):

$$M + e^- \rightarrow M_{\bullet}^+ + 2e^- \tag{8.1}$$

This is an inefficient process; out of 1 million molecules, on average, only one molecule will be ionized. In spite of this, enough ions are normally produced in the ion source, and GC-MS is thus a very sensitive technique that can be used to detect very low concentration levels.

Under EI, $M_{\bullet}^+$ is formed, which is called the molecular ion. This has the same mass as the original molecule, because the mass of the single electron that was removed is negligible. The molecular ion is positively charged because one electron was removed during the ionization. In addition to the charge, the symbol for the molecular ions is usually supplied with a small dot to indicate that they are *radical ions*. This means that the molecular ion contains one unpaired electron. Normally, all the electrons in a molecule are paired, but when one electron is removed, one of the remaining electrons will be unpaired.

Generally, most molecular ions formed by EI are unstable. Many of them will immediately decompose to form fragment ions. In cases where virtually all molecular ions decompose to form fragment ions, no molecular ions will be found in the corresponding mass spectrum. An example of this is shown in Figure 8.16 (upper spectrum). In the mass spectrum of amphetamine (central stimulant), there is no peak at mass 135 corresponding to the molecular mass.

In other cases, molecular ions are more stable and will appear in the EI mass spectrum. In such cases, the molecular mass of the compound can be read directly from the mass spectrum. The molecular mass will correspond to one of the highest masses observed in the mass spectrum. An example of this is shown in Figure 8.16 (lower spectrum). The mass spectrum of lysergide (lysergic acid diethylamide (LSD), a hallucinogen) contains a strong

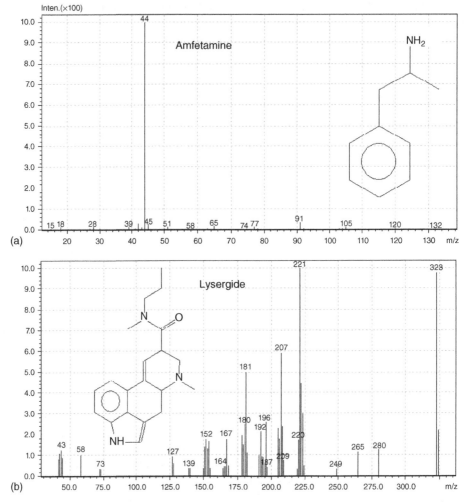

Figure 8.16 *Mass spectrum (electron ionization) of (a) amphetamine and (b) lysergide (lysergic acid diethylamide, or LSD)*

signal at mass 323 corresponding to the molecular ion. For most organic compounds, the majority of the molecular ions will decompose to fragment ions under EI conditions.

8.10.3   A Closer Look into EI Mass Spectra

Atomic masses are measured in *atomic mass units* (amu), and 1 amu is defined as 1/12 of the mass of the ^{12}C-isotope. This means that the mass of the ^{12}C-isotope is exactly 12.0000 amu. The mass of other atoms are measured relative to the mass of the ^{12}C-isotope. The ^1H-isotope, for example, is 11.9068 times lighter than the ^{12}C-isotope, and the mass is therefore:

$$(1/11.9068) \cdot 12.0000 \text{ amu} = 1.0078 \text{ amu}$$

The ^{35}Cl-isotope is 2.91407 times heavier than the ^{12}C-isotope, and the mass is therefore:

$$2.91407 \cdot 12.0000 \text{ amu} = 34.9688 \text{ amu}$$

The mass of a certain molecule is calculated by the sum of the masses of the individual isotopes. The mass of chlorambucil $^{12}C_{14}{}^{1}H_{19}{}^{35}Cl_{2}{}^{14}N^{16}O_{2}$ (which contains the isotopes ^{12}C, ^{1}H, ^{35}Cl, ^{14}N, and ^{16}O) is therefore:

$$(14 \cdot 12.0000 + 19 \cdot 1.0078 + 2 \cdot 34.9688 + 14.0031 + 2 \cdot 15.9949) \text{ amu}$$
$$= 303.0787 \text{ amu}$$

In GC-MS (and LC-MS), masses are measured in amu or in *Daltons*, which is the same as amu (1 amu = 1 Da). Atomic and molecular masses as given here with 3–4 decimals are known as *exact masses*. Mass spectrometry with *high resolution* measures exact masses, whereas in mass spectrometry with *low resolution*, the masses are rounded to the nearest integer. Masses rounded to the nearest integer are called *nominal masses*. Measured in nominal masses, the molecular mass of chlorambucil $^{12}C_{14}{}^{1}H_{19}{}^{35}Cl_{2}{}^{14}N^{16}O_{2}$ is therefore:

$$(14 \cdot 12 + 19 \cdot 1 + 2 \cdot 35 + 14 + 2 \cdot 16) \text{ amu} = 303 \text{ amu}$$

As most routine GC-MS is performed with low resolution, we will principally use nominal masses in this chapter.

As mentioned in Section 8.10.1, mass spectra are bar plots where the quantities (intensities) of the molecular ion and the fragment ions are plotted as a function of the mass-to-charge ratio (m/z). In a mass spectrum, we can therefore see the fragment ions formed in the ion source, and in what relative quantities they are formed. In the magnified mass spectrum for chlorambucil (Figure 8.17), for example, we see a large peak for mass 254. This tells us that this particular fragment, corresponding to M-CH$_2$Cl$^+$, is formed in large quantities.

In addition to this main fragment ion, a large number of other ions also occur, but these are formed in substantially smaller quantities (lower intensity). One example is at mass 268 corresponding to (M-Cl)$^+$. Most fragment ions from low-molecular compounds acquire only one charge ($z = 1$) during EI, and the m/z values are therefore practically equal to the mass of the fragments (m/z = m). Because of this, we can assume the mass of the main fragment ion for chlorambucil is 254 and the mass of the molecular ion is 303. In low-resolution mass spectrometry, bars are plotted at nominal masses. This means that all exact masses such as those recorded between 253.5000 and 254.5000 Da will be plotted as a single bar with mass 254.

The y-axis in a mass spectrum shows the amount or intensity of the different ions. A peak or a bar with high intensity thus corresponds to the formation of many ions with that particular mass. The intensity of the different ions will therefore depend on the amount of sample that enters the mass spectrometer; high amounts of sample in the ion source will provide high amounts of fragment ions. To make mass spectra independent of concentration, it is common to scale the spectra relative to the most intense fragment ion. The most intense fragment ion is called the *base peak*, and the intensity of the base peak is scaled to 100%. The other fragment ions are then scaled relative to the base peak.

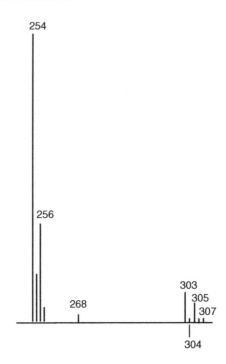

Figure 8.17   *Magnified mass spectrum for chlorambucil in the mass range of 250–310*

Unlike the other methods that have been discussed in this textbook, mass spectrometry differentiates between different isotopes, and this is important to take into consideration when interpreting mass spectra. With chlorine as an example, there are two naturally occurring isotopes: one is ^{35}Cl isotope with mass 34.968855 ($\approx$35), and the other isotope is ^{37}Cl, which has a mass of 36.965896 ($\approx$37). The isotope with the lowest mass is dominant and constitutes 75.77%, whereas the heavier isotope constitutes 24.23%. Because these two isotopes have different masses, the mass spectrum for chlorambucil ($C_{14}H_{19}Cl_2NO_2$), which contains two chlorine atoms, will have three peaks for the molecular ion (Figure 8.17); mass 303 corresponds to molecular ions with two ^{35}Cl atoms, mass 305 contains one ^{35}Cl and one ^{37}Cl, and mass 307 contains two ^{37}Cl atoms. Mass 303 is the highest signal as it is most probable that both Cl atoms are ^{35}Cl. The probability for this is $0.75 \cdot 0.75 = 0.56$. Mass 305 is lower in intensity than mass 303 because it is less probable that one of the Cl atoms is ^{37}Cl. The probability for this is $0.75 \cdot 0.25 + 0.25 \cdot 0.75 = 0.38$, and the intensity of 305 is $(0.38/0.56) \cdot 100\% = 68\%$ of mass 303. Mass 307 is even lower because the probability for two ^{37}Cl atoms in the same molecular ion is relatively low. The probability for this is $0.25 \cdot 0.25 = 0.06$, and mass 307 corresponds to $(0.06/0.56) \cdot 100\% = 11\%$ of mass 303.

It is not only chlorine that naturally occurs with multiple isotopes. In practice, all elements frequently found in organic compounds exist as several isotopes except for fluorine, iodine, and phosphorus. Table 8.1 gives a list of the different isotopes and their natural occurrence (%).

As seen in Table 8.1, carbon naturally exists as two isotopes, namely, the ^{12}C-isotope with mass 12.00000 ($\approx$12) and the ^{13}C-isotope with mass 13.003354 ($\approx$13). The

Table 8.1   *Masses and occurrence of stable isotopes*

Isotope	Mass	Occurrence (%)
^1H	1.007825	99.985
^2H	2.014102	0.015
^{12}C	12.00000	98.90
^{13}C	13.003354	1.10
^{14}N	14.003074	99.634
^{15}N	15.000108	0.366
^{16}O	15.994915	99.762
^{17}O	16.999133	0.038
^{18}O	17.999160	0.200
^{19}F	18.998405	100
^{28}Si	27.976927	92.23
^{29}Si	28.976491	4.67
^{30}Si	29.973761	3.10
^{31}P	30.973763	100
^{32}S	31.972074	95.02
^{33}S	32.971461	0.75
^{34}S	33.967865	4.21
^{36}S	35.967091	0.02
^{35}Cl	34.968855	75.77
^{37}Cl	36.965896	24.231
^{79}Br	78.918348	50.69
^{81}Br	80.916344	49.31
^{127}I	126.904352	100

^{12}C-isotope dominates and constitutes 98.90%, whereas the ^{13}C-isotope represents only 1.10%. In the molecular ion for chlorambucil, there are a total of 14 carbon atoms, and $14 \cdot 1.1\% = 15.4\%$ of these molecular ions will contain one ^{13}C-atom. This is the reason why a significant peak at mass 304 can be observed (see Figure 8.17). For chlorambucil, the peak at mass 254 corresponds to the loss of CH_2Cl from the molecular ion $(C_{14}H_{19}Cl_2NO_2 - CH_2Cl = C_{13}H_{17}ClNO_2)$. This fragment ion contains 13 carbon atoms, and an average of $13 \cdot 1.1\% = 14.3\%$ of the ions with the formula $C_{13}H_{17}ClNO_2$ will contain one ^{13}C-atom. Therefore, a significant peak is observed at mass 255 (see Figure 8.17). As discussed later, the isotope ratios are important in connection with the interpretation of unknown spectra, and they can provide significant structural information.

8.10.4   GC-MS Identification Using EI Spectra

EI mass spectra of a given substance acquired with GC-MS in different laboratories with different instruments are normally very similar when EI is used at 70 eV. Because of this, mass spectra are ideal for comparison, and an unknown compound can often be identified by comparing its mass spectrum with a *reference spectrum* for a known compound. If the spectrum of the unknown compound is identical that of to the reference spectrum, the substance can be identified tentatively. If this is combined with retention time identification, the certainty of the identification is improved substantially.

Today, mass spectra at 70 eV have been recorded and published for a very large number of different drug substances, drug metabolites, and other chemical compounds. These spectra are stored in computer-based libraries. GC-MS systems are delivered with such libraries connected online to the instrument. After recording a mass spectrum of a given unknown compound, the bioanalytical chemist can electronically search for reference spectra in the database. This makes it possible to make a quick and easy verification of the analyte. Libraries containing mass spectra for more than 100 000 different organic compounds are commercially available, and these are a very important tool for identification. Bioanalytical measurements are normally performed for analytes which previously have been characterized by mass spectrometry, and reference spectra are available and ready for use. An example of the use of a reference spectrum for identification of an unknown substance is shown in Box 8.2.

Box 8.2 Gas Chromatography–Mass Spectrometry (GC-MS) Identification of Fencamfamin in Urine Sample (Upper Spectrum) Using the Reference Spectrum (Lower Spectrum) Recorded with Electron Ionization at 70 eV

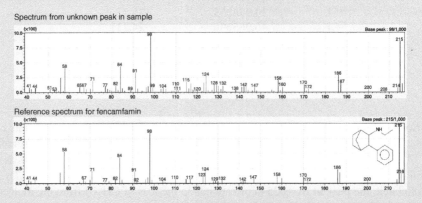

The electronic search for reference spectra will often give a list of several suggestions that can match the spectrum of the unknown compound. The list will be ranked according to the fit quality between the spectrum of the unknown and the spectrum of the reference. The comparison is based on the 5–10 most intense ions. The suggestions should be controlled by the operator to verify that all major fragments in the reference spectrum are also present in the spectrum of the unknown substance. In this case, the unknown can be tentatively identified from the reference spectrum, but the identification should be further supported by retention times. Thus, a pure standard solution of the compound should be injected, and the retention time of this should be similar to the retention time of the unknown.

Occasionally, unknown compounds in GC-MS cannot be identified because no matching spectra are found in the libraries. In such cases, the bioanalytical chemist may interpret the mass spectrum manually to get some structural information (characterization). Interpretation of mass spectra is complex and requires detailed knowledge of fragmentation patterns of organic compounds. This is beyond the scope of this textbook, but some of the principles of interpretation will be discussed related to a few examples, to give a flavor of the field and a better understanding of EI mass spectra.

For manual interpretation of mass spectra, it is common to start from the upper part of the mass scale, because the fragments with the highest masses are those most specific for the structure. First, the molecular ion $(M)^+$ should be located to possibly get the molecular mass of the unknown. With EI, there are no signals at higher masses than the molecular ion, and therefore the molecular ion should be at the right-hand side of the mass spectrum. The highest mass will usually not directly equal the molecular mass, but be shifted one or two mass units $((M+1)^+$ or $(M+2)^+)$ upward from the molecular ion because of ^{13}C-isotopes and other heavier isotopes.

Figure 8.18 shows an example of how to determine the molecular mass. The highest mass with a signal is at 286. This signal corresponds to molecular ions containing one ^{13}C-atom $(M+1)^+$, whereas the signal at mass 285 corresponds to the actual molecular mass of the compound $(M)^+$. There is also a peak at mass 284, but this arises from loss of an H-radical from the molecular ion $(M-1)^+$.

Ions with lower masses than the molecular ion have been formed by fragmentations from the molecular ion. These fragmentations can provide significant structural information. Table 8.2 gives different possible fragmentations directly from molecular ions, the corresponding mass differences, and possible structural information that can be derived from the actual mass shifts. Please note that Table 8.2 is not complete. For interpretational work, a more complete list should be used. This can be found in a number of specialized textbooks on MS.

Use of the information that the different isotopes provide is very important in mass spectral interpretation, as discussed in Chapter 7. With very clean mass spectra without interferences from other substances, the ^{13}C-isotope can be used to estimate the number of carbon atoms in the molecular ion as well as in fragment ions. If the intensity of an ion with mass m is I_m, and the intensity of the corresponding ion with one ^{13}C-atom (mass $m+1$)

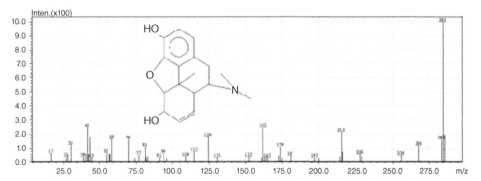

Figure 8.18 *Mass spectrum (electron ionization) of morphine (an analgesic drug). The signal at mass 286 corresponds to molecular ions with one ^{13}C-atom $(M+1)^+$, and 285 corresponds to the actual molecular mass of the compound $(M)^+$*

Table 8.2   *Typical fragmentations from the molecular ion*

Mass	Loss	Interpretation
M-1	H	
M-2	H_2	
M-15	CH_3	α-Cleavage
M-16	O	Aromatic nitro compound, N-oxide
M-16	NH_2	Primary amide
M-17	OH	Carboxylic acid, tertiary alcohol
M-18	H_2O	Primary alcohol
M-19	F	Fluoride
M-20	HF	Fluoride
M-26	C_2H_2	Unsubstituted aromatic hydrocarbon
M-27	C_2H_3	Ethyl ester
M-28	C_2H_4	Aromatic ethyl ether, n-propylketone
M-28	CO	
M-29	C_2H_5	Ethyl ketone, α-cleavage
M-29	HCO	Aliphatic aldehyde
M-30	CH_2O	Aromatic methyl ether
M-30	NO	Aromatic nitro compound
M-31	OCH_3	Methyl ester, aromatic methyl ester
M-32	CH_3OH	Methyl ether
M-33	SH	Thiol
M-34	H_2S	Thiol
M-35/37	Cl	Chlorinated compound
M-36/38	HCl	Chlorinated aliphatic compound
M-41	C_3H_5	Propyl ester
M-42	CH_2CO	Aromatic acetate, $ArNHCOCH_3$
M-42	C_3H_6	Aromatic propyl ether
M-43	C_3H_7	Propyl ketone, α-cleavage
M-43	CH_3CO	Methyl ketone
M-45	OC_2H_5	Ethyl ester
M-45	COOH	Carboxylic acid
M-46	NO_2	Aromatic nitro compound
M-46	C_2H_5OH	Ethyl ester
M-60	CH_3CO_2H	Acetate
M-127	I	Iodated compound

is I_{m+1}, the following equation is valid:

$$I_{m+1} = n \cdot 0.011 \cdot I_m \tag{8.2}$$

where n is the total number of carbon atoms in the ion. This number can then be determined by the following rearrangement of Equation 8.2:

$$n = \frac{I_{m+1}}{0.011 \cdot I_m} \tag{8.3}$$

An example of how to calculate the number of carbon atoms is shown in Box 8.3.

Box 8.3 Calculation of the Number of Carbon Atoms Based on the Ratio between ^{12}C and ^{13}C

From the mass spectrum of chlorambucil (Figure 8.15), the following intensities are measured for the molecular ion with only ^{12}C-isotopes (I_m) and the molecular ion with one ^{13}C-isotope (I_{m+1}):

$$I_m = 10.7\%$$

$$I_{m+1} = 1.6\%$$

This intensity ratio corresponds to the following number of carbon atoms:

$$n = \frac{1.6\%}{0.011 \cdot 10.7\%} = 13.6 \approx 14$$

The calculation demonstrates that, within the uncertainty of the measurement, the molecular ion contains 14 carbon atoms.

In addition to the ^{13}C-isotope, isotope patterns of chlorine and bromine are also important when interpreting mass spectra in bioanalysis. As shown in Table 8.1, chlorine as ^{35}Cl and ^{37}Cl is in the ratio $\approx 3:1$, and bromine as ^{79}Br and ^{81}Br is in the ratio $\approx 1:1$. The isotope patterns of these are therefore very characteristic, as illustrated in Figure 8.19, and all molecular ions and fragment ions containing Cl or Br are relatively easy to locate (see, e.g., the mass spectrum for chlorambucil in Figure 8.17).

To give a flavor of manual interpretation, a few mass spectra (EI) will be discussed in this section. Attention will be focused on only direct decompositions from the molecular ion, whereas further fragmentation products will not be discussed in this textbook.

Figure 8.20 shows the mass spectrum for p-amino-benzoic acid. For this compound, a strong signal for the molecular ion is observed at mass 137. This information tells that the molecular ion is relatively stable, and the m/z value of the molecular ion gives the molecular mass of the compound. Note that the molecular ion has an odd nominal mass. This follows the *nitrogen rule*, which states that if a compound has an odd number of nitrogen atoms,

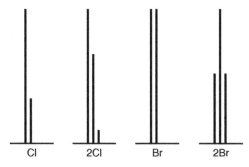

Figure 8.19 *Isotope patterns of chlorine (one and two atoms) and bromine (one and two atoms). Two mass units between each peak*

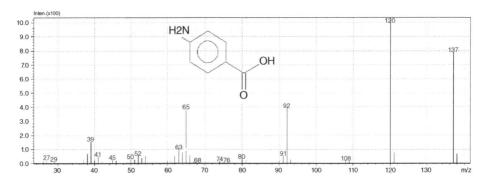

Figure 8.20   *Mass spectrum (electron ionization) of p-amino-benzoic acid*

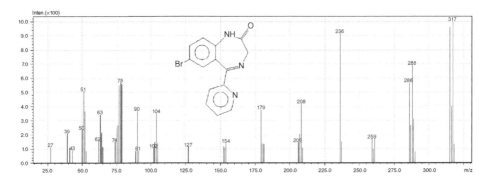

Figure 8.21   *Mass spectrum (electron ionization) of bromazepam*

then the molecular ion has an odd nominal mass. If a compound has an even number of nitrogen atoms or no nitrogen atoms, the molecular ion has an even nominal mass.

The next fragment of interest is found at mass 120, which makes a mass difference of 17 compared to the molecular ion. According to Table 8.2, this corresponds to the loss of an OH radical $(M-17)^+$, which is frequently observed for carboxylic acids and tertiary alcohols. We also find a $(M-45)^+$ fragment at mass 92 corresponding to the loss of a COOH radical, and this confirms that the compound is a carboxylic acid.

A number of drugs contain halogen atoms (F, Cl, Br, and I). According to Table 8.2, such compounds may give characteristic release of the halogens corresponding to $(M-19)^+$ for F, $(M-35)^+$ for Cl, $(M-79)^+$ for Br, and $(M-127)^+$ for I. An example of the release of bromine is shown in Figure 8.21 for bromazepam (a tranquilizer).

The molecular ion for bromazepam is found at mass 315. This is consistent with the nitrogen rule, because bromazepam contains three nitrogen atoms. The compound contains a bromine atom, and the typical isotope distribution of bromine is found in the mass spectrum (compare with Figure 8.19); the heavier bromine isotope ^{81}Br gives a peak at mass 317 with about the same intensity as the peak at 315. Normally, bromine is easily released from the molecular ion, as a C–Br bond in general is relatively weak compared to C–F and

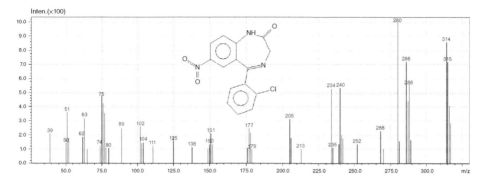

Figure 8.22   *Mass spectrum (electron ionization) of clonazepam*

C–Cl bonds. For bromazepam, a peak at $(M-79)^+$ is observed, corresponding to mass 236. The fragment ion at mass 236 no longer contains bromine, and no bromine isotope pattern is observed at this fragment. Mass 236 is a large peak in the spectrum that confirms that the release of bromine from the molecular ion in this case is a major fragmentation path. The isotope pattern at the molecular ion and the $(M-79)^+$ decomposition tell clearly that this compound contains bromine.

An example of the release of chlorine is shown in Figure 8.22 for clonazepam (an anti-convulsant). Clonazepam contains three nitrogen atoms, and the molecular ion is located at mass 315. Clonazepam contains one chlorine atom, and a signal with one-third of the intensity of mass 315 is observed at mass 317, which corresponds to the ^{37}Cl isotope. This gives a strong indication of the presence of chlorine. In addition, a strong signal is observed at m/z 280, and this signal is due to the loss of chlorine from the molecular ion. Observe that the chlorine isotope pattern is not present at m/z 280. This supports that a single chlorine atom is present in the structure.

A strong signal is also observed at m/z 314, and this corresponds to the loss of a hydrogen radical from the molecular ion. A strong signal at mass 286 is observed, and this corresponds to the loss of CO from m/z 314.

8.10.5   GC with Chemical Ionization (CI)

As we have seen in Section 8.10.2, the molecular ion may be unstable during electron ionization in GC-MS, and it may totally decompose to fragment ions. In such cases, the molecular ion will not be present in the mass spectrum. For identification, the lack of molecular ions is a significant disadvantage. If EI does not provide sufficient molecular ions, it means that the electrons that bombard the sample contain too much energy. The high energy is causing all the molecular ions to form fragment ions. This can be circumvented by reducing the energy of the electrons, and this is done by adding a *reagent gas* into the ion source. This is known as *chemical ionization* (CI). Methane is a typical reagent gas used for CI. The reagent gas is supplied in relatively large amounts compared to the amount of sample (1000–10 000 times more reagent gas than the sample). Electrons released by the rhenium or tungsten filament in the ion source will therefore almost exclusively ionize the reagent gas. With methane as reagent gas, CH_5^+ and $C_2H_5^+$ will be formed. These ions are highly reactive and will react with the analyte (XH) according to the following schemes:

$$CH_5^+ + XH \rightarrow XH_2^+ + CH_4 \tag{8.4}$$

$$C_2H_5^+ + XH \rightarrow XH_2^+ + C_2H_4 \tag{8.5}$$

$$C_2H_5^+ + XH \rightarrow X^+ + C_2H_6 \tag{8.6}$$

In Equations 8.4 and 8.5, the CI gives a *pseudo-molecular ion* with mass $(M+1)^+$. In the lower reaction (Equation 8.6), the CI gives a pseudo-molecular ion with mass $(M-1)^+$. Thus, CI is accomplished with ions, whereas EI is accomplished with electrons.

CI with methane normally forms ions with either mass $(M+1)^+$ or $(M-1)^+$. These ions are relatively stable, and therefore they appear as intense signals in the mass spectra. Because of this, CI is very powerful for the determination of molecular masses. An example of this is illustrated in Figure 8.23, with the mass spectra for amphetamine both with EI and with CI.

In the EI spectrum of amphetamine, it is clear that all the molecular ions have fragmented, mainly to $C_2H_6N^+$ with mass 44, and there is no signal detected at mass 135 corresponding to the molecular mass. Using CI with methane, however, the base peak is at mass 136. This corresponds to $(M+1)^+$, and the molecular mass of the compound may be determined

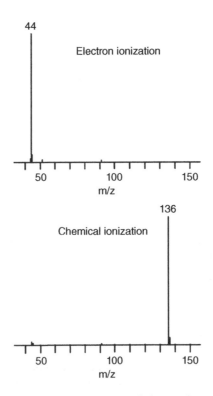

Figure 8.23 *Mass spectra (electron ionization and chemical ionization) of amphetamine*

to be 135. Generally, high mass signals (like m/z 136 in Figure 8.23) are much more specific for the chemical structure than low mass signals (like m/z 44 in Figure 8.23), and for identification purposes, high mass signals are preferred.

CI is usually carried out with methane as the reagent gas. In addition to methane, other gases such as iso-butane and ammonia can be used. Iso-butane and ammonia are used in cases where less fragmentation is requested. These reagents bind H^+ more strongly than CH_4 does and impart less energy to MH^+ when the proton is transferred to M. In addition to this, both positive and negative ions can be produced. Compounds containing electronegative atoms form significant amounts of negative ions, and these ions may be analyzed with high sensitivity and specificity by *negative chemical ionization mass spectrometry.*

8.10.6 GC-MS with High-Resolution Mass Spectrometry

In some cases, the use of nominal masses is insufficient because important structural information can be lost. An example of this is shown in Box 8.4: a molecular ion with nominal mass 120 can correspond to several different types of ions that mutually have very small differences in their exact mass. The elemental compositions of the various ions at mass 120 are very different, and for correct interpretation of the mass spectrum, it may be essential to know the correct elemental composition. Elemental compositions can be determined if the mass determinations are highly accurate. This is done by mass spectrometry with high resolution, where the mass spectrometer is tuned for highly accurate mass determinations. If the mass of the molecular ion in Box 8.4 is determined to be 120 070 ($\pm$0005), it can be concluded that this corresponds to the elemental composition of $C_7H_8N_2$.

Box 8.4 Several Ions with the Same Nominal Mass but with Significant Differences in Exact Mass

Ion	Nominal mass	Exact mass
$C_5H_4N_4^+$	120	120.044
$C_7H_8N_2^+$	120	120.069
$C_9H_{12}^+$	120	120.096

High-resolution mass spectrometry can thus determine the elemental composition of both molecular ions and fragment ions.

High-resolution mass spectrometry requires highly sophisticated instrumentation, and not all mass spectrometers can be used for accurate mass determinations.

8.10.7 GC-MS Instrumentation and Modes of Operation

MS is performed under vacuum conditions, with a very low pressure inside the instrument as discussed in Chapter 7. The reason for this is to prevent the ions of interest from colliding with molecules from the air. This ensures that the ions reach the detector. GC-MS is done with capillary columns, and the outlet of the capillary column is placed directly inside the ion source of the mass spectrometer. Thus, EI or CI takes place directly when the compounds leave the GC column. The ion source is a small chamber inside the MS system just before the mass analyzer. Usually, 1–3 ml/min of carrier gas (helium or hydrogen) is used in capillary GC, and this gas will also enter the mass spectrometer. To maintain a low pressure in the mass spectrometer, the instrument must be closed, and air penetrating the system must constantly be pumped out of the system by means of powerful pumps. The capacity of the pumps that evacuate the mass spectrometer is so powerful that the pressure does not rise significantly due to the carrier gas coming from the GC column. The pumps are connected to the ion source, the mass analyzer, and the detector, and they ensure that the pressure inside the instrument does not exceed 10^{-4} to 10^{-8} Torr (10^{-7}–10^{-11} bar).

After ionization, the ions pass through a very narrow opening into the mass spectrometer and further into the mass analyzer and the detector. In the mass analyzer, the ions are separated according to their m/z values. Many GC-MS instruments in use today are using a quadrupole mass analyzer. This mass analyzer has already been discussed in Chapter 7. In addition to quadrupole mass analyzers, ion trap analyzers and time-of-flight analyzers also are used frequently in GC-MS. Also, these systems have been presented in Chapter 7.

In GC-MS, the mass spectrometer can be operated in different modes, similar to what was discussed for LC-MS:

- Full scan and recording of spectra (full scan)
- Selected ion monitoring (SIM)
- Selected reaction monitoring (SRM).

To maximize structural information, it is common to operate the mass spectrometer in the *full scan mode*, where the instrument continuously records full mass spectra in a given mass range. Because the separated substances are in the mass spectrometer only for a very short time in GC-MS, it is important that mass spectra are recorded with very short time intervals. Thus, typically 1–5 mass spectra are recorded per second. The mass spectra are stored during the analysis in a computer, and the results can be plotted as a chromatogram (detector signal versus retention time) showing a peak for each of the separated and ionized substances. The separated substances can be identified or characterized by examining their mass spectra stored in the computer. The total amount of ions (total ion current, or TIC), or the sum of all ions, in each mass spectrum is plotted continuously as a function of separation time, resulting in a *TIC chromatogram* (see Figure 8.24).

The TIC is very low when only the carrier gas is flowing into the mass spectrometer. Each time a compound is eluting from the GC and into the mass spectrometer, the TIC increases for a short period of time, giving rise to a peak in the TIC chromatogram. By recording the mass spectra, a lot of structural information is available for each of the separated substances, but the drawback is that sensitivity is limited in the full scan mode, and it may therefore be difficult to detect compounds at very low concentration levels.

For trace-level measurements and for quantitative measurements, it is common to operate the GC-MS system in the *SIM mode*. As discussed in Chapter 7, the mass spectrometer is tuned to measure only a single mass or a few selected masses during the chromatographic separation. From SIM analysis, a *selected ion chromatogram* is obtained, which is

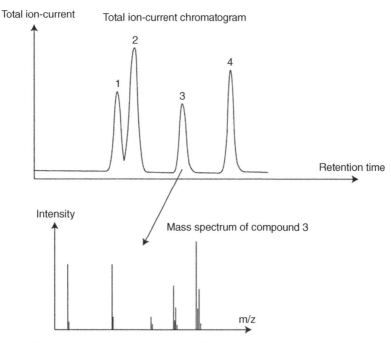

Figure 8.24 *Total ion current chromatogram (TIC) and mass spectrum for component 3 in a mixture of four components*

a plot of the intensity of the particular mass as a function of retention time. Components in the sample that give the requested mass during ionization and fragmentation will therefore be found as peaks in the selected ion chromatogram (see Figure 8.25), whereas other compounds that do not form ions with that particular mass will not be displayed.

If SIM does not give adequate sensitivity or specificity, one can take advantage of the SRM. SRM analysis can be accomplished with a *triple quadrupole mass spectrometer*, as discussed in Chapter 7. Such measurements can be made with very high sensitivity and specificity because ion background from other substances is effectively eliminated.

8.10.8 Quantitative GC-MS

As we have discussed, GC-MS is very suitable for identification. However, GC-MS can also be used for quantitative analysis. Quantitative analysis is based on the fact that areas under an analyte peak in a selected ion chromatogram (or a SRM chromatogram) are proportional to the concentration of the analyte. By calibration with standard solutions of known concentration, one can plot the standard curves of peak area as a function of concentration, and from these one can determine the amount of the substance in unknown samples. This is similar to quantification by other chromatography detectors, as discussed in Chapter 5. To quantify N-desmethylclobazam in plasma using SIM at mass 286 (see also Figure 8.25),

Ion signal m/z 286

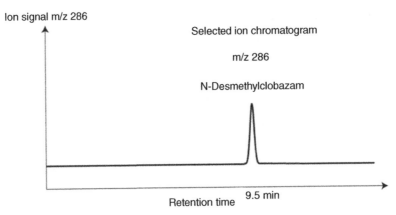

Figure 8.25 *Selected ion chromatogram (mass 286) for gas chromatography–mass spectrometry (GC-MS) analysis of human plasma. N-desmethylclobazam eluted into the mass spectrometer at 9.5 minutes (retention time) and formed ions with mass 286, resulting in the peak observed in the selected ion chromatogram. Plasma samples contain a large number of other substances, but these do not appear because the ionization or fragmentation does not form ions with mass 286*

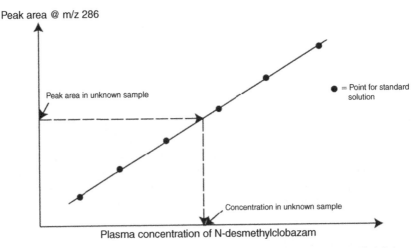

Figure 8.26 *Calibration curve for the quantitative determination of N-desmethylclobazam in plasma based on six standard solutions with different concentrations of N-desmethylclobazam*

a calibration curve has to be established. Known amounts of N-desmethylclobazam are added to drug-free plasma and analyzed by GC-MS. Peak area in the selected ion chromatogram for mass 286 is plotted as a function of concentration (Figure 8.26). The standard curve can be used for GC-MS of unknown plasma samples to determine concentrations of the drug substance. Quantification using GC-MS with SIM is widely used in bioanalysis.

9

Analysis of Small-Molecule Drugs in Biological Fluids

Steen Honoré Hansen[1] and Stig Pedersen-Bjergaard[2]

[1]*School of Pharmaceutical Sciences, University of Copenhagen, Denmark*
[2]*School of Pharmacy, University of Oslo, Norway*
School of Pharmaceutical Sciences, University of Copenhagen, Denmark

In the first part of this textbook (Chapters 1–8), the different techniques in use for bioanalysis have been presented, with special focus on the basic principles. All this you have to read, understand, and remember before you read Chapters 9 and 10. Chapter 9 is intended to give selected examples on how to use the different techniques with a focus on small-molecule drugs. The examples include analysis from plasma and serum samples, whole blood samples, dried blood spots, urine samples, and saliva. The examples have been carefully selected to represent typical applications, and each example is provided with detailed questions and explanations related to the experimental procedures. Hopefully, this will give the reader the understanding of all the small details that are difficult to get from more general textbooks in analytical chemistry. You should not try remembering all the details in the procedures from Chapters 9 and 10, but rather focus on the understanding. Bioanalytical methods vary a lot, but if you understand the examples given in Chapters 9 and 10, you have a very good platform for understanding most of the methods in use.

9.1 Plasma and Serum Samples

Analysis for drug substances and their metabolites in plasma or serum is the standard way to obtain knowledge about the drug exposure in a test animal or a human being. The data obtained show how much of the drug has reached the blood circulation. In most bioanalytical methods, it is the total amount of drug substance or metabolite that is measured, although it is generally accepted that it is only the free fraction of the drug in plasma that

Bioanalysis of Pharmaceuticals: Sample Preparation, Separation Techniques and Mass Spectrometry,
First Edition. Steen Honoré Hansen and Stig Pedersen-Bjergaard.
© 2015 John Wiley & Sons, Ltd. Published 2015 by John Wiley & Sons, Ltd.

is able to exhibit its activity. If knowledge of the free fraction is needed, this has to be separated from the protein-bound fraction by, for example, equilibrium dialysis or ultrafiltration. For most purposes, plasma and serum behave equally, but there is a small difference as described in Chapter 3. If very small samples are handled, it has to be considered if the anticoagulant addition plays a role with plasma. In that case serum should be used, but in many cases with larger animals and humans this will be of no concern.

The purpose of analysis of plasma and serum samples is to obtain knowledge about the concentration of drug substances and their metabolites in the circulating blood system. Different examples related to analysis of plasma and serum samples are discussed in Boxes 9.1–9.5.

Box 9.1 Fast LC-MS Analysis of Polar Analytes Using Silica HPLC Columns

Purpose and Procedure

The current procedure is intended to be used for identification and quantitative determination of morphine derivatives and their metabolites in plasma or serum samples. The method is to be used in clinical trials in different investigations of pain treatments. The procedure is based on solid phase extraction (SPE) as sample preparation and liquid chromatography–mass spectrometry (LC-MS/MS) for separation and quantification.

Target Analytes

Morphine, morphine-3-glucuronide (M3G), and morphine-6-glucuronide (M6G). Chemical structure and physicochemical properties of the analytes:

	Morphine	M3G	M6G	Hydromorphone
log D pH = 7.4	−0.07	−1.12	−0.79	0.15
pKa	8.2 (amine)	2.7 (carboxylic acid)	2.9 (carboxylic acid)	8.1 (amine)
	9.9 (phenol)	8.2 (amine)	9.1 (phenol)	9.5 (phenol)

At neutral pH, morphine and hydromorphone are basic substances, whereas M3G and M6G are amphoteric.

Internal Standard

Two possibilities were used: (i) hydromorphone or (ii) deuterated (D$_3$) standards of morphine, M3G, and M6G prepared in water. Hydromorphone is suitable, but the deuterated standards are preferred for the final method.

Sample Collection and Pretreatment

Samples of blood are collected in a Vacutainer™ tube containing EDTA. After thorough mixing, the vials are centrifuged at 2000 g for 15 minutes, and the plasma is transferred to another vial and frozen at −20 °C until analysis.

Sample Preparation

SPE using Oasis hydrophilic-lipophilic balance (HLB) 10 mg 96-well extraction plates. After the samples have been thawed and thoroughly mixed, they are centrifuged at 2000 g for 5 minutes.

Two hundred and fifty microliters are transferred to a 1-ml, 96-well deep well plate by a multipipette. Between two consecutive pipettings, the multipipette needles are washed with water, and with 0.5% trifluoroacetic acid (TFA) in acetonitrile and water. Twenty-five microliters of internal standard spiking solution (50/100/1000 ng/ml of morphine-d$_3$/M6G-d$_3$/M3G-d$_3$) are added to all samples, calibration standards, and quality control (QC) samples. After thorough mixing, the samples are treated by SPE.

The 10 mg Oasis HLB 96-well plate is first conditioned using 0.8 ml of methanol, followed by 0.8 ml 0.1% TFA in water. Samples are mixed with 250 μl of 0.1% TFA in water and then loaded to the cartridge plate. After the samples have passed through the SPE cartridges, the plate is washed with 0.8 ml 0.1% TFA. The plate is then eluted with two portions of 0.4 ml of methanol–water (1:1, v/v) into a deep well collection plate. The collection plate is dried by vacuum centrifugation (in a TurboVap™) and reconstituted in 150 μl of acetonitrile–water–TFA (95:5:0.01, v/v/v). Finally, the collection plate is heat-sealed with a film and placed in the LC-MS/MS system for analysis. All handling is performed using robots.

HPLC-MS/MS

LC-MS/MS is performed under the following conditions:

Column	Silica, 50 mm × 3.0 mm inside diameter (I.D.); 5 μm particles
Mobile phase	Acetonitrile–water–TFA (90:10:0.01, v/v/v)
Flow rate	1 ml/min
Column temperature	Ambient
Injection volume	15 μl
MS polarity	Positive

ESI	2.0 kV	
MS/MS transitions	Morphine	$286 \rightarrow 152$ (d_3 : $289 \rightarrow 152$)
	M6G	$462 \rightarrow 286$ (d_3 : $465 \rightarrow 289$)
	M3G	$462 \rightarrow 286$ (d_3 : $465 \rightarrow 289$)
Calibration range	Morphine	0.5–50 ng/ml plasma
	M6G	2.0–200 ng/ml plasma
	M3G	2.0–2000 ng/ml plasma
QC samples	Morphine/M6G/M3G: 2:5:50; 10:25:250; 40:100:1000 ng/ml plasma	
LLOQ	Morphine/M6G/M3G: 0.5/2/2 ng/ml plasma	

Identification

The identification of the analytes in the plasma samples is based on a combination of the retention time and the MS/MS transition data. In this case, the MS/MS data are identical for M6G and M3G, and therefore separation in the LC system is of major importance (Figure 9.1). The identification of each of the two glucuronides therefore has to be confirmed with authentic reference standards.

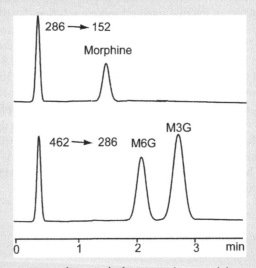

Figure 9.1 *Chromatogram of a sample from a patient receiving a dose of morphine. Mass spectrometry (MS) detection is performed by MS/MS.*

Calibration and Quantitative Analysis

Quantification of analytes using high-performance liquid chromatography (HPLC) has to be performed relative to known reference standards. For each analyte, a calibration curve is obtained from a number of calibration standard solutions. The calibration standard solutions are made up in water or in the actual matrix (in this case, in blank plasma obtained from healthy, drug-free persons). It is preferable to use calibration standard solutions prepared in the actual matrix and to treat these solutions in an identical procedure as the unknown samples.

The concentration of morphine in an unknown sample is determined using a calibration curve:

Calibration standard solution	Morphine (ng/ml)	Morphine-d3 ng added	Peak area $286 \rightarrow 152$	Peak area $289 \rightarrow 152$	Peak area ratio
1	0.5	1.25	210	2005	0.11
2	2	1.25	809	1989	0.41
3	5	1.25	2011	1951	1.03
4	15	1.25	5989	1910	3.14
5	25	1.25	9902	1995	4.96
6	50	1.25	20047	2010	9.97
Unknown sample					
Rep. 1	X	1.25	1588	1998	0.79
Rep. 2	X	1.25	1532	1973	0.78

Linear regression results in an equation: $y = 0.1988x + 0.0397$, where y is the measured peak ratio. Using the average of the two repetitions of the unknown sample:

$$0.79 = 0.1988x + 0.0397$$

$$x = 3.8 \text{ ng morphine/ml plasma.}$$

Calculation of the concentration of the other two analytes is performed in a similar way.

This example is based on Shou WZ *et al.*[1].

Fundamental Questions to the Procedure: Step by Step

First try to answer the following questions by yourself, and subsequently read our discussion below:

1. Why is this both an identification method and a quantification method?
2. What is a Vacutainer® tube?
3. Why does it contain EDTA?
4. Why are the vials centrifuged at 2000 g for 15 minutes?
5. Why are the samples stored at $-20\,°C$?
6. Why is the sample volume 250 µl?
7. Why is an internal standard added to the samples and calibration solutions?
8. Why are the deuterated standards preferred over hydromorphone?

9. Why is TFA used in the sample preparation?
10. Why is SPE used for sample preparation?
11. Why is the SPE column washed with 0.1% TFA after application of sample?
12. Why are the analytes eluted from the SPE using methanol–water (1:1, v/v)?
13. Why is the eluate evaporated to dryness?
14. Why is the evaporated sample reconstituted in acetonitrile–water–TFA (95:5:0.01, v/v/v)?
15. Why is a silica HPLC column used for the separation?
16. Why is the mobile phase composed of acetonitrile–water–TFA (90:10:0.01, v/v/v)?
17. Why is MS/MS used for detection?
18. Why is the ESI used in positive mode?
19. Why does the calibration range differ from analyte to analyte?
20. Why do the analytes show different lower limits of quantification (LLOQs)?

Discussion of the Questions

1. The purpose of this method is for monitoring in clinical trials where quantitative data are needed. It is important that the quantitative data are reliable, and therefore it is important to make sure that the peak measured also has the correct identity.
2. Vacutainer® is a collection tube for whole blood. When sampling, a sterile needle is connected to the tube through a septum, and the vacuum inside the tube will suck whole blood into the tube.
3. To prevent the sampled blood from coagulation, EDTA is present in the Vacutainer®. EDTA binds the Ca^{2+} needed for coagulation, and therefore no coagulation takes place.
4. The Vacutainer® is then centrifuged in order to separate the blood cells and blood plates from the plasma. After centrifugation, the yellow clear plasma is transferred to another tube.
5. Samples are normally stored at $-20\,°C$ unless a lower temperature ($-80\,°C$) is required. This is to improve sample stability. Often, samples are collected over a period of time and then analyzed. It is important that the analytes are stable during the storage period. When the sample has been thawed, it is important to secure a thorough mixing because some inhomogeneity has occurred during storage.
6. In general, samples as small as possible are used, especially when repeated samples from a person are needed (to follow the concentration over time). The sample volume used depends on the required LLOQ and the sensitivity of the detector.
7. The internal standards are added to the sample as early in the sample preparation procedure as possible in order to improve repeatability and accuracy. This is especially important when using detectors where the experimental condition may vary over time.

8. When an internal standard is used for each analyte, different behavior that may result in different recovery of the analytes will be compensated for each analyte. If another compound like hydromorphone is used as internal standard, it has to be separated from the other analytes, and it may behave a little differently compared to the analytes to be quantified.

9. TFA is used to control pH in the sample. The glucuronides are tertiary amines as well as carboxylic acid. The pH is kept low with TFA and the carboxylic acid groups will be unionized, in this way improving retention on the SPE cartridge. Furthermore, the change in pH in the plasma sample will also help to break possible protein bindings.

10. SPE is used to partially purify the sample. Plasma proteins and a lot of inorganic ions will be eliminated as they pass through the column during sample application.

11. The following wash with 0.1% TFA will further improve the removal of proteins and inorganic ions.

12. The analytes are eluted with a solvent sufficiently strong enough. If a higher concentration of methanol is used, more of the other more hydrophobic plasma constituents will also be eluted.

13. The eluate is evaporated as it cannot be injected directly into the HPLC (50% methanol is too strong of an eluent in the system used) and also in order to concentrate the analytes.

14. The evaporated sample is reconstituted in a solvent that is less strongly eluting compared to the mobile phase. This is because water is the strongest solvent in this type of chromatography.

15. The principle of chromatography used in hydrophilic interaction liquid chromatography (HILIC) is a kind of normal phase (straight phase) chromatography where some water is a part of the mobile phase. The principle is especially suited for polar compounds and especially for compounds with an amine structure. Morphine and in particular glucuronides are polar substances. The most polar analytes are retained the most.

16. The HPLC system is an isocratic system using a mobile phase containing 90% of acetonitrile as this has been shown to give a satisfactory retention and separation. If the acetonitrile content is increased, the retention will increase.

17. MS/MS is used to obtain as high a selectivity as possible. This will improve the reliability of the data obtained. In this case, it is very important to be sure that no overlap is present in the isolation of m/z 286 and m/z 289 for morphine and morphine-d3, respectively, because the fragmentation gives the same m/z 152 fragment.

18. The analytes will be positively charged when they elute from the HPLC column and will therefore easily be detected using positive ESI.

19. The difference in the calibration range for the analytes reflects the expected concentration range to be found in the unknown plasma samples from the clinical trials.

20. A difference in LLOQ of the analytes is observed, and this is due to the difference in their chemical structure. At neutral pH, morphine is a tertiary amine whereas the glucuronides are zwitterionic in nature, but in the HPLC system used all analytes are present as protonated amines. However, the glucuronides are much more polar than morphine itself, and they also have a higher molecular mass.

Box 9.2 Enantiomeric Separation and Quantification of Verapamil and Its Active Metabolite, Norverapamil, in Human Plasma by LC-MS/MS

Purpose and Procedure

The current procedure is intended to be used for quantitative determination of verapamil and its metabolite norverapamil in a pharmacokinetic study on healthy humans. Verapamil is dosed as the racemic mixture, and a chiral method provides the possibility to investigate possible enantiomeric discrimination in the ADME (adsorption, distribution, metabolism, and excretion) process of the drug and its metabolite.

The procedure is based on a liquid–liquid extraction (LLE) as sample preparation and LC-MS/MS for chiral separation and quantification.

Target Analytes

R-verapamil, S-verapamil, R-norverapamil, and S-norverapamil.
Chemical structure and physicochemical properties of some of the analytes:

R, S-Verapamil   R, S-Norverapamil

	R-verapamil	S-verapamil	R-norverapamil	S-norverapamil
log D pH = 7.4	2.7	2.7	1.35	1.35
pKa	9.7	9.7	9.8	9.8

Internal Standard

R,S-verapamil-d_6 and R,S-norverapamil-d_6

Sample Collection and Pretreatment

Samples of blood are collected in lithium heparin-vacuettes. After thorough mixing, the vials are centrifuged at 3000 g for 10 minutes, and the plasma is transferred to another vial and frozen at −20 °C until analysis.

Sample Preparation

LLE with a mixture of n-hexane and diethyl ether (50:50, v/v). After the samples have been thawed and thoroughly mixed, they are centrifuged at 2000 g for 5 minutes.

Fifty microliters of plasma are mixed with 50 µl of internal standard (0.5 µg/ml of each of R,S-verapamil-d_6 and R,S-norverapamil-d_6 in methanol). To this mixture, 2.0 ml of the extraction solvent is added and the extraction is performed by rotation (30 rpm for 5 minutes). The vial is then centrifuged at 3200 g for 10 minutes. The supernatant is transferred to another vial and evaporated at 40 °C under nitrogen. The dried sample is reconstituted in 100 µl of acetonitrile–0.05% TFA (30:70, v/v), and 5 µl is injected into the LC.

HPLC-MS/MS

LC-MS/MS is performed under the following conditions:

Column	Chiracel OD-RH, 150 mm × 4.6 mm I.D.; 5 µm particles	
Mobile phase	Acetonitrile–0.05% TFA (30:70, v/v)	
Flow rate	0.6 ml/min with flow splitting 1:1	
Column temperature	40 °C	
Injection volume	5 µl	
MS polarity	Positive	
ESI	4.0 kV	
MS/MS transitions	R- and S-verapamil	455.3 → 165.1 (d_6 : 461.3 → 165.1)
	R- and S-norverapamil	441.3 → 165.1 (d_6 : 447.3 → 165.1)
Calibration range	R- and S-verapamil	1.0–250 ng/ml plasma
	R- and S-norverapamil	1.0–250 ng/ml plasma
QC samples	R,S-verapamil and R,S-norverapamil at three concentration levels: 3.0, 100, and 200 ng/ml plasma	
LLOQ	1.0 ng/ml for all four analytes	

Identification

The identification of the analytes in the plasma samples is based on a combination of the retention time and the MS/MS transition data. In this case, the MS/MS data are identical for the R- and S-forms of the two analytes, and therefore separation of all four analytes in the LC system is of major importance if any change in enantiomeric ratio is to be verified. The identification of each of the enantiomers is also to be confirmed with authentic reference standards of the optical isomers.

Stability of Samples

Stability of the samples is investigated using spiked plasma standards. Long-term stability of spiked samples stored at −20 and −80 °C is studied over the necessary period of time (months). Freeze–thaw stability is evaluated using at least three freeze–thaw cycles.

Short-term stability of the final sample preparation to be injected is studied over 24 hours in the auto-sampler.

Calibration and Quantitative Analysis

Quantification of analytes using HPLC has to be performed relative to known reference standards. For each analyte, a calibration curve is obtained from a number of calibration standard solutions. The calibration standard solutions are made up in water or in the actual matrix (in this case, in blank plasma obtained from healthy, drug-free persons). It is preferable to use calibration standard solutions prepared in the actual matrix and to treat these solutions in an identical procedure as the unknown samples.

The calculations of the drug concentration of each analyte are performed identical to what has been shown in Box 9.1.

This example is based on Singhal P *et al.* [2].

Fundamental Questions to the Procedure—Step by Step

First try to answer the following questions by yourself, and subsequently read our discussion below:

1. Why are both identification and quantification parts of this method?
2. What is a lithium heparin-vacuette?
3. Why does it contain lithium heparin?
4. Why is the sample volume 50 µl?
5. Why is internal standard added to the sample?
6. How does the internal standard differ from the analyte?
7. Why is LLE used?
8. Why are n-hexane and diethyl ether used for the extraction?
9. Why is the mixture centrifuged after extraction?
10. Why is the evaporation performed at 40 °C using nitrogen?
11. Why is the dried sample reconstituted in 100 µl?
12. Why is the dried sample reconstituted in acetonitrile–0.05% TFA (30:70, v/v)?
13. Why is only 5 µl injected into the LC?
14. Why is a Chiralcel OD-RH Column used?
15. Why is a mobile phase used that is composed of acetonitrile and 0.05% TFA?
16. Why is the column heated to 40 °C?
17. Why is MS/MS operated in positive mode?
18. Why is the ESI voltage set at 4 kV?
19. Why are all four analytes measured at the fragment m/z 165.1?
20. Why is stability evaluated by using freeze–thaw cycles?
21. Why is long-term stability investigated at −20 and −80 °C?

Discussion of the Questions

1. The purpose of the present method is to be used for a pharmacokinetic study of the analytes after dosing to humans. It is important that the quantitative data obtained on each isomer are reliable, and therefore it is important to make sure that the peak measured also has the correct identity.

2. A vacuette is similar to a Vacutainer®—a vial with a vacuum to ease the collection of the blood.

3. The heparin is added to the vacuette to prevent the blood from coagulation.

4. The sample volume is kept as low as possible for the sake of the test persons, especially when repeated sample from a person is needed (to follow the concentration over time). The sample volume used is dependent on the required LLOQ and the sensitivity of the detector.

5. The internal standards are added to the sample as early in the sample preparation procedure as possible in order to improve repeatability and accuracy. This is especially important when using detectors where the experimental condition may vary over time, and this is sometimes the case in MS where especially the spray efficiency can vary.

6. The internal standards in this case all have replaced six hydrogen for six deuterium, and the m/z value will therefore increase with six mass units compared to the analyte itself. There is therefore no overlap with isotope peaks.

7. LLE often gives cleaner extracts compared to SPE.

8. The mixture of n-hexane and diethyl ether provides a suitable polarity. Using n-hexane alone will probably result in lower recovery. The mixture is also sufficiently volatile to be removed by evaporation. In contrast, n-hexane is a solvent that should be avoided due to its toxicity. It may be replaced with n-heptane.

9. The aqueous phase in this solvent extraction procedure is relatively small, some of the phase may adhere to the walls of the vial, and some emulsion may be formed. Centrifugation secures an efficient separation of the two liquid phases.

10. When solvent evaporates, energy is consumed and the temperature of the remaining solvent will decrease and thus reduce the vapor pressure and evaporation. When adding heat, the vapor pressure will increase; and the nitrogen helps to remove the vapors from the surface, making the evaporation more effective.

11. The reconstitution is made in as little solvent volume as possible, but one also has to be able to handle it. After the evaporation, the sample is distributed over the inner surface in the bottom of the vial. To be able to dissolve all, a certain volume is needed.

12. The reconstituted liquid is identical to the mobile phase into which the sample is to be injected. This is a standard procedure to avoid any compatibility problems.

13. The injection volume is kept low (5 µl) because chiral stationary phases should not be overloaded. Overloading will result in a decrease in resolution due to peak broadening.

14. The Chiralcel OD-RH column contains a stationary phase made of cellulose tris(3,5-dimethylphenylcarbamate) coated on 5 µm silica–gel particles. The

cellulose is chiral in nature due to the glucose units. If or when two enantiomers have different affinity to the stationary phase, they can be separated. In this case, good separation is achieved.

15. The retention and separation depend on the composition of the mobile phase. The optimal composition has to be found by experiment as it depends on the nature of the analytes.

16. Also, the optimal column temperature has to be found be experiment. In general, column efficiency increases with column temperature due to better phase transfer (van Deemter equation).

17. The MS/MS is operated in positive mode because the analytes are amines that are protonated in the mobile phase. The analytes therefore already are in the ionic state (cationic) when eluted from the analytical column.

18. By experiment, the MS/MS is optimized, including the ion spray voltage, to give the highest signal as possible.

19. All four analytes are fragmented to the same fragment, including the internal standards. Thus, the fragment containing the tertiary amine and the secondary amine functionality is not part of the fragment used for the final detection.

20. Sometimes samples are thawed and then frozen again. During this procedure, the analytes may degrade, and therefore this is often evaluated to assure stability.

21. Even if samples are frozen at $-80\,^\circ$C, chemical reactions may take place. The only way to obtain knowledge about this is to perform experiments.

Box 9.3 Fast Determination of Antiepileptic Drugs in Serum Using Protein Precipitation and UHPLC-ESI Tandem MS/MS

Purpose and Procedure

The current procedure is intended to be used for quantitative determination of selected antiepileptic drug substances in serum to be used for therapeutic drug monitoring (TDM).

The procedure is based on precipitation of the proteins with an organic solvent, followed by ultrahigh performance LC-MS (UHPLC-MS/MS) separation and quantification.

Target Analytes

Lacosamide, lamotrigine, levetiracetam, primidone, and zonisamide.

Chemical structure and physicochemical properties of the analytes:

Lacosamide   Lamotrigine   Levetiracetam   Primidone   Zonisamide

	Lacosamide	Lamotrigine	Levetiracetam	Primidone	Zonisamide
log P	−0.02	1.93	−0.6	1.12	0.11
pKa	—	5.9	—	—	9.8

Internal Standards

Lacosamide-d_3, lamotrigine-$^{13}C_3d_3$, levitiracetam-d_3, primidone-d_5, and zonisamide-d_4.

Sample Collection and Pretreatment

Samples of blood are collected in S-Monovette serum collection devices. The monovette is thoroughly mixed and left for 30 minutes. After centrifugation at 3000 g for 10 minutes, the serum is transferred to another vial and frozen at −20 °C until analysis.

Sample Preparation

Protein precipitation with acetonitrile and zinc sulfate. After the samples have been thawed and thoroughly mixed, they are centrifuged at 2000 g for 5 minutes.

Fifty microliters of serum are added to 200 µl of 0.1 M zinc sulfate and 500 µl of acetonitrile containing the internal standards (1.5 µg/ml of each of the five internal standards). After thorough mixing, the mixture is centrifuged at 14 000 g for 3 minutes. The supernatant is transferred to an auto-sampler vial, and 1 µl is injected into the LC.

UHPLC-MS/MS

LC-MS/MS is performed under the following conditions:

Column	Acquity UPLC BEH Phenyl, 50 mm × 2.1 mm I.D.; 1.7 µm particles
Mobile phases	A: 0.1% formic acid with 2 mM ammonium formate added B: 0.1% formic acid in methanol with 2 mM ammonium formate added
Gradient	0–1.0 minute 90% A; 1.0–1.8 minutes 90 to 5% A; 1.8–2.3 minutes 5% A; 2.5 minutes 90% A
Flow rate	0.5 ml/min
Column temperature	60 °C
Injection volume	1 µl
MS polarity	Positive
ESI	1.0 kV

MS/MS transitions	Lacosamide	$251.0 \rightarrow 108.0$ ($d_3 : 254.0 \rightarrow 108.0$)
	Lamotrigine	$256.0 \rightarrow 145.0$ ($^{13}C_3, d_3 : 262.0 \rightarrow 148.0$)
	Levetiracetam	$171.1 \rightarrow 126.1$ ($d_3 : 174.1 \rightarrow 129.1$)
	Primidone	$219.1 \rightarrow 162.1$ ($d_5 : 224.1 \rightarrow 167.1$)
	Zonisamide	$213.1 \rightarrow 132.0$ ($d_4 : 217.1 \rightarrow 136.0$)
Calibration range	For each of the five analytes	$0.1–40.0\,\mu g/ml$ serum
QC samples	All five analytes at three concentration levels	$0.2, 2.5,$ and $20\,\mu g/ml$ serum
LLOQ	$0.1\,\mu g/ml$ serum for all five analytes	

Identification

The identification of the analytes in the serum samples is based on a combination of the retention time and the MS/MS transition data. In this case, the MS/MS data are most important because some of the analytes have very similar retention times in the LC system. The identification of each analyte is also confirmed with reference to the coeluting internal standard.

Stability of Samples

QC samples are stored at $-20, 4\,°C$, and room temperature for one month, one week, and one day, respectively, and then analyzed. Freeze–thaw stability is evaluated using at least three freeze–thaw cycles.

Short-term stability of the final sample preparation to be injected is studied over 24 hours in the auto-sampler.

Calibration and Quantitative Analysis

Quantification of analytes using HPLC has to be performed relative to known reference standards. For each analyte, a calibration curve is obtained from a number of calibration standard solutions. The calibration standard solutions are made up in water or in the actual matrix (in this case, in blank plasma obtained from healthy, drug-free persons). It is preferable to use calibration standard solutions prepared in the actual matrix and to treat these solutions in an identical procedure as the unknown samples.

The calculations of the drug concentration of each analyte are performed identical to what has been shown in Box 9.1.

Ion Suppression

Ion suppression and enhancement effects are studied by post column injection of solution of the drugs (10 μg/ml) and injecting blank serum samples after sample preparation.

This example is based on reference [3].

Fundamental Questions to the Procedure—Step by Step

First try to answer the following questions by yourself, and subsequently read our discussion below:

1. What is TDM?
2. Why is serum used for the analysis?
3. Why are both identification and quantification parts of this method?
4. What is protein precipitation?
5. Why are proteins removed from the serum samples?
6. Why are the internal standards made to have an m/z value 3–6 amu higher than the drug substance itself?
7. What is an S-monovette serum collection device?
8. Why is zinc sulfate added in the sample preparation?
9. Why is 500 μl acetonitrile added in the sample preparation?
10. Why are the samples centrifuged after they have been thawed?
11. Why is the sample mixture after protein precipitation centrifuged at 14 000 g?
12. Why is a short UPLC column used for the separation?
13. Why is gradient elution used in this method?
14. Why is the flow rate 0.5 ml/min?
15. Why is the column kept at 60 °C?
16. Is it relevant to test the stability of the samples at −20 °C for one month considering the purpose of this method?
17. What is ion suppression/enhancement?

Discussion of the Questions

1. *TDM* stands for *therapeutic drug monitoring*, and it is performed when it is important to verify that a given dose of the drug results in a therapeutic concentration of the drug substance in the blood of the patient.
2. In this method, serum is used for the determination of the drug concentration, but in principle plasma also could have been used. The only difference between plasma and serum is that the fibrin/fibrinogen system for coagulation has been removed from the plasma. The two types of bioliquids are very similar in behavior and will normally result in identical analysis results.
3. Because the quantitative data are to be used for drug monitoring, it is important to be sure that the peak measured in the chromatogram is correctly identified. This is primarily done using the MS/MS signal, which is very selective. The

retention of the peak measured also has to be identical to the retention time of the reference standard.

4. Protein precipitation happens when the solubility of the proteins in the matrix (plasma or serum) is decreased. The solubility will decrease if a high concentration of salt is added, if strong acid is added, if certain divalent metal ions are added, or if organic solvents are added.

5. Proteins have to be removed almost completely from plasma or serum before chromatographic analysis because otherwise the proteins may precipitate when injected into the mobile phase in the LC system and in this way clog the column.

6. The internal standards are synthesized to have at least 3 amu higher than the drug substance itself to avoid any overlap with isotope mass (e.g., 13C isotope mass) in the mass spectrum from the drug substance itself.

7. An S-monovette serum collection device is a sealed vial provided with a coagulation activator, which allows for complete coagulation of the blood within 20–30 minutes. The coagulation activator is coated onto plastic granules, which form a layer between the coagulated blood and serum sample after centrifugation.

8. Zinc ions will ease the protein precipitation.

9. The addition of 500 µl of acetonitrile to the 250 µl mixture of serum and zinc sulfate solution will precipitate the proteins, and the presence of zinc ions will further improve this process. A mixture of plasma/serum and acetonitrile in a ratio of 1:2 is normally sufficient to precipitate the proteins.

10. When thawing samples after storage, they may contain some turbidity or minor precipitate, which should be removed.

11. Centrifugation at 14 000 g after protein precipitation will remove any particles from the solution, which then can be injected into the LC system without risk of clogging the column.

12. The UPLC–UHPLC column contains column packing materials with very small particle diameter. They therefore also result in higher back pressure. The small particle diameter provides high separation efficiency, and therefore a shorter column may be used. At the same time, the very selective MS/MS detection principle is used, which allows identification and quantification of the analytes even if they are not fully separated in the LC system.

13. Furthermore, gradient elution is used to even further shorten the analysis time. The gradient helps to elute otherwise late-eluting peaks faster due to the increase in eluting strength of the gradient.

14. The flow rate is set at 0.5 ml/min, which is in accordance with the knowledge that the optimal efficiency of UHPLC columns is at a higher linear flow rate compared to columns with larger particle sizes.

15. The column is kept at 60 °C, where the mass transfer and thus the column efficiency are improved. In this way, the separation efficiency is optimized. At the same time, the viscosity of the mobile phase is decreased, which will result in a lower back pressure.

16. The present method is intended to be used for TDM, and the answer about the concentration of the drug substance in the blood is used to regulate the dose given to the patient. Collected blood samples will therefore not be stored in the freezer but will be analyzed within a short time after collection.

17. Ion suppression or enhancement is a phenomenon often seen when using MS detection. It is due to intermolecular interactions taking place in the ionization chamber where the eluent from the LC column is sprayed out and evaporated. One way to evaluate this is to infuse a solution of the analyte after the separation column into the eluent coming from the LC column and before the spray chamber. This constant infusion will result in a constant signal, and if a blank serum sample (which has been prepared according to the method) injected into the LC system results in a deflection from the constant signal, this is a sign of ion suppression (lower signal) or enhancement (higher signal). If this deflection takes place with the same retention time as that of the analyte, the quantitative data obtained will be biased.

Box 9.4 Fast Analysis of Morphine and Its Two Glucuronide Metabolites Using UHPLC-MS/MS

Purpose and Procedure

The current procedure is intended to be used for identification and quantitative determination of morphine and its two major metabolites in plasma. The method is to be used in clinical trials in different investigations of pain treatments. The procedure is based on protein precipitation/SPE for phospholipid removal as sample preparation and LC-MS/MS for separation and quantification.

Target Analytes

Morphine, M3G, and M6G.

Chemical structure and physicochemical properties of the analytes:

Morphine   M3G   M6G   Hydromorphone

	Morphine	M3G	M6G	Hydromorphone
Monoisotopic mass	285.1365	461.1686	461.1686	285.1365
log D pH = 7.4	−0.07	−1.12	−0.79	0.15
pKa	8.2 (amine)	2.7 (carboxylic acid)	2.9 (carboxylic acid)	8.1 (amine)
	9.9 (phenol)	8.2 (amine)	9.1 (phenol)	9.5 (phenol)

At neutral pH, morphine and hydromorphone are basic substances, whereas M3G and M6G are amphoteric.

Internal Standard

Hydromorphone.

Sample Collection and Pretreatment

Samples of blood are collected in Vacutainer® plastic K_2EDTA tubes. After gently mixing by inversion about 10 minutes, the vials are centrifuged at 1500 g for 10 minutes, and the plasma is transferred to another vial and frozen at −80 °C until analysis.

Sample Preparation

Protein precipitation is performed in Phree® 96-well SPE plates. After the samples have been thawed and thoroughly mixed, they are centrifuged at 2000 g for 5 minutes.

First 360 µl acetonitrile is applied to each well, then 10 µl of internal standard (0.5 µg hydromorphone/ml) is added, and finally 180 µl of plasma sample is added. Mixing is performed on a Vortex-genie 2 for 2 minutes, and then the 96-well plate is centrifuged at 500 g for 5 minutes into a 96-well collection plate. The filtrate is evaporated in a vacuum centrifuge at 45 °C for approximately 90 minutes, and the residue is reconstituted in 100 µl mobile phase A. Finally, the collection plate is heat-sealed with a film and placed in the LC-MS/MS system for analysis.

UHPLC-MS/MS

LC-MS/MS is performed under the following conditions:

Column	Kinetex XB-C18, 100 Å, 50 mm × 2.1 mm I.D.; 1.7 µm solid core particles
Mobile phase	A: 1% acetonitrile containing 15 mM ammonium formate (pH 3.5)
	B: 90% acetonitrile containing 15 mM ammonium formate (pH 3.5)

Gradient	0 minute 95% A; 0–1.7 minutes 90 to 70% A; 1.8 minutes 95% A; re-equilibration time, 1.3 minutes
Flow rate	0.5 ml/min
Column temperature	Ambient
Injection volume	15 μl
MS polarity	Positive
ESI	2.0 kV
MS/MS transitions	Morphine 286.1 → 181.1 Hydromorphone 286.1 → 201.1 M6G and M3G 462.1 → 286.1
Calibration range	Morphine 0.1–32 ng/ml plasma M6G 1.0–200 ng/ml plasma M3G 1.0–400 ng/ml plasma
QC samples	Morphine/M6G/M3G: 2:5:50; 10:25:250; 40:100:1000 ng/ml plasma
LLOQ	Morphine/M6G/M3G: 0.5/1/1 ng/ml plasma

Identification

The identification of the analytes in the plasma samples is based on a combination of the retention time and the MS/MS transition data. In this case, the MS/MS data are identical for M6G and M3G, and therefore separation in the LC system is of major importance (Figure 9.2). Also, the mass of morphine and hydromorphone is identical,

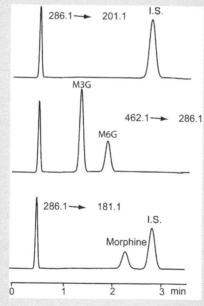

Figure 9.2 *Chromatograms of a sample from a patient receiving a dose of 10 mg morphine. Detection is performed using mass spectrometry (MS)/MS.*

and even if two different transitions are used hydromorphone also has a fragment at 181.1. Therefore, these two compounds have to be separated in the chromatographic system. The identification of each of the two glucuronides also has to be confirmed with authentic reference standards because their MS/MS signals are identical.

Calibration and Quantitative Analysis

Quantification of analytes using HPLC has to be performed relative to known reference standards. For each analyte, a calibration curve is obtained from a number of calibration standard solutions. The calibration standard solutions are made up in the actual matrix (in this case, in blank plasma obtained from healthy, drug-free persons). It is preferable to use calibration standard solutions prepared in the actual matrix and to treat these solutions in an identical procedure as the unknown samples.

The concentration of morphine in an unknown sample is determined using a calibration curve:

Calibration standard solution	Morphine (ng/ml)	Hydromorphone ng added	Peak area 286.1 → 181.1	Peak area IS 286.1 → 201.1	Peak area ratio
1	1.0	5.0	490	11 216	0.041
2	4.0	5.0	1 909	11 029	0.173
3	8.0	5.0	3 511	11 055	0.318
4	16	5.0	7 489	10 997	0.681
5	24	5.0	10 989	11 230	0.978
6	32	5.0	14 752	11 013	1.340
Unknown sample					
Rep. 1	X	5.0	5 556	11 339	0.490
Rep. 2	X	5.0	5 521	11 173	0.494

Linear regression results in an equation: $y = 0.041623x - 0.001155$, where y is the measured peak ratio. Using the average of the two repetitions of the unknown sample:

$$0.492 = 0.041623x + 0.001155$$

$$x = 11.8 \text{ ng morphine/ml plasma.}$$

Calculation of the concentration of the two glucuronides is performed in a similar way.

(This example is based on Charlotte Gabel-Jensen and Steen Honoré Hansen, unpublished data.)

Fundamental Questions to the Procedure—Step by Step

First try to answer the following questions by yourself, and subsequently read our discussion below:

1. Why is this both an identification method and a quantification method?
2. What is a Vacutainer® plastic tube?
3. Why does it contain K_2EDTA?
4. Why are the vials mixed by inversion for 10 minutes?
5. Why are the vials centrifuged at 1500 g for 10 minutes?
6. Why are the samples stored at −80 °C?
7. Why are the samples mixed and centrifuged at 2000 g for 5 minutes after having been thawed?
8. Why is acetonitrile added to the Phree® plate first?
9. Why is an internal standard added to the samples and calibration solutions?
10. Why is the sample volume 180 μl?
11. Why is the Phree 96-well plate treated in a Vortex-genie for 2 minutes?
12. Why is the plate centrifuged at only 500 g for 5 minutes?
13. Why is the filtrate evaporated?
14. Why is the residue reconstituted in 100 μl mobile phase A?
15. Why is the collection plate sealed with a film?
16. What is the chromatographic separation principle?
17. Why are 1.7 μm solid core particles used?
18. Why is gradient elution used in the separation?
19. Why is the flow rate 0.5 ml/min?
20. Why is 15 μl of the sample injected?
21. Why is MS/MS used for detection?
22. Why is the ESI used in positive mode?
23. Why does the calibration range differ from analyte to analyte?
24. Why do the analytes show different LLOQs?

Discussion of the Questions

1. The purpose of this method is for monitoring clinical trials where quantitative data are needed. It is important that the quantitative data are reliable, and therefore it is important to make sure that the peak measured also has the correct identity.
2. Vacutainer® is a collection tube for whole blood. When sampling, a sterile needle is connected to the tube through a septum, and the vacuum inside the tube will suck whole blood into the tube.
3. To prevent the sampled blood from coagulation, EDTA is present (as the potassium salt) in the Vacutainer®. EDTA binds the Ca^{2+} needed for coagulation, and therefore no coagulation takes place.
4. It is important to make sure that the collected plasma is mixed with the K_2EDTA.

5. The Vacutainer® is then centrifuged in order to separate the blood cells and blood plates from the plasma. After centrifugation, the yellow clear plasma is transferred to another tube.

6. Samples are normally stored at −20 °C unless a lower temperature (−80 °C) is required. This is to improve sample stability. Often, samples are collected over a period of time and then analyzed. It is important that the analytes are stable during the storage period. When the sample has been thawed, it is important to secure a thorough mixing because some inhomogeneity has occurred during storage.

7. During the freezing and thawing processes, some fractionation of the sample may take place. To obtain a homogeneous sample, mixing is important in order to be able to sample a representative aliquot.

8. The acetonitrile is added to the Phree® plate first in order to prevent the sample from contact with the SPE before the plasma proteins have been precipitated. The Phree® plate will retain the analytes as well as phospholipids; the phospholipids will be permanently retained.

9. The internal standards are added to the sample and calibration standards as early in the sample preparation procedure as possible in order to improve repeatability and accuracy. This is especially important when using detectors where the experimental condition may vary over time.

10. The sample volume is 180 μl because the aim is to obtain a precipitation mixture with a ratio between sample and acetonitrile of 1:2. This will result in full precipitation of the proteins.

11. A treatment in the Vortex-genie for 2 minutes is to ensure thorough mixing between sample and acetonitrile.

12. When centrifuged, the precipitated proteins are retained in the 96-well plate, and the SPE material will concomitantly retain phospholipids.

13. The filtrate contains about 66% acetonitrile. The initial conditions in the LC system contain about 5% acetonitrile. The 66% acetonitrile will therefore be too strongly eluting when injected. Therefore, it is necessary to evaporate the filtrate and re-dissolve it in mobile A.

14. In order to use a final sample for injection that is as concentrated as possible, the evaporated filtrate is re-dissolved in only 100 μl mobile phase A.

15. The 96-well plate is covered with a film to prevent evaporation from the many samples before injection into the LC system.

16. The chromatographic separation principle is reversed phase chromatography, and it is performed using gradient elution.

17. The small particle diameter (1.7 μm) of the stationary phase provides higher separation efficiency per unit length of the column, and therefore a short column can be used. This again will provide reduced time of analysis.

18. Gradient elution is used to obtain a faster elution of late-eluting peaks, including peaks that otherwise might interfere with the analytes in the next chromatogram.

19. The flow rate of 0.5 ml/min is higher than the normal standard 0.2–0.25 ml/min for 2.1 mm I.D. columns. However, the small particle diameter prevents any loss in efficiency due to the higher linear flow rate, and this will also add to a higher throughput.
20. Fifteen microliters is injected in order to inject as much as possible without losing separation efficiency. The internal diameter of the column is only 2.1 mm.
21. MS/MS is used to obtain as high selectivity as possible. This will improve the reliability of the data obtained.
22. The analytes will all be positively charged when they elute from the UHPLC column and will therefore easily be detected using positive ESI.
23. The difference in the calibration range for the analytes reflects the expected concentration range to be found in the unknown plasma samples from the clinical trials.
24. A difference in LLOQ of the analytes is observed, and this is due to the difference in their chemical structure. Morphine is present as a protonated tertiary amine, whereas the glucuronides are zwitterionic at pH 3.5.

Box 9.5 Rapid, Simultaneous Determination of Lopinavir and Ritonavir in Human Plasma Using Protein Precipitation and Salting Out LLE Prior to Ultrafast LC-MS/MS

Purpose and Procedure

The current procedure is intended to be used for rapid quantitative determination of lopinavir and ritonavir in human plasma. The method is to be used in clinical trials. The procedure is based on protein precipitation with organic solvent and zinc sulfate. A salting-out effect further improves the sample preparation to be used in LC-MS/MS for separation and quantification.

Target Analytes

Lopinavir and ritonavir.
   Chemical structure and physicochemical properties of the two analytes:

Lopinavir

Ritonavir

	Lopinavir	Ritonavir
Monoisotopic mass	628.3824	720.3128
log P	3.9	4.2
pKa	—	—

Lopinavir and ritonavir are neutral substances.

Internal Standard

Deuterated analogs.

Sample Collection and Pretreatment

Samples of blood are collected in Vacutainer^{TM} plastic K_2EDTA tubes. After gently mixing by inversion about 10 minutes, the vials are centrifuged at 1500 g for 10 minutes and the plasma is transferred to another vial and frozen at $-80\,°C$ until analysis.

Sample Preparation

Samples are thawed completely in room temperature water and mixed thoroughly. Sample preparation is performed in 96-well plates.

To 50 μl internal standard working solution (500 ng ml^{-1} each of lopinavir-d$_5$ and ritonavir-d$_5$ in 50% acetonitrile), add 50 μl sample and mix thoroughly (aspirating and dispensing 100 μl five times). Then add 200 μl acetonitrile and 100 μl 3 M zinc sulfate and mix thoroughly (aspirating and dispensing 300 μl five times). Centrifuge the 96-well plate at 3000 g for 4 minutes at 10 °C. Transfer 100 μl of the organic layer (the upper layer) into a new 96-well plate, and add 100 μl of water. After thorough mixing, inject 10 μl into the LC-MS/MS system.

HPLC-MS/MS

LC-MS/MS is performed under the following conditions:

Column	Zorbax Extend-C18 rapid resolution HT, 30 mm × 2.1 mm I.D.; with 1.8 μm particles
Mobile phase	Acetonitrile and water (55:45, v/v) containing 0.1% formic acid
Flow rate	0.5 ml/min
Column temperature	Ambient
Injection volume	10 μl
MS polarity	Positive
ESI	2.0 kV

MS data collection	From 8 to 48 seconds	
MS/MS transitions	Lopinavir	$629 \rightarrow 447$ (d_5 : $634 \rightarrow 452$)
	Ritonavir	$721 \rightarrow 296$ (d_5 : $726 \rightarrow 296$)
Calibration range	Lopivanir	0.02–16.0 μg/ml plasma
	Ritonavir	0.01–8.00 μg/ml plasma
QC samples	Lopinavir	0.04, 1.04, and 13.0 μg/ml plasma
	Ritonavir	0.02, 0.52, and 6.50 μg/ml plasma
LLOQ	Lopinavir	0.02 μg/ml plasma
	Ritonavir	0.01 μg/ml plasma

Identification

The identification of the analytes in the plasma samples is based on a combination of the retention time and the MS/MS transition data.

Please note: The time of chromatography per sample is only 1 minute (see MS data collection).

Calibration and Quantitative Analysis

Quantification of analytes using HPLC has to be performed relative to known reference standards. For each analyte, a calibration curve is obtained from a number of calibration standard solutions. The calibration standard solutions are made up in the actual matrix (in this case, in blank plasma obtained from healthy, drug-free persons). It is preferable to use calibration standard solutions prepared in the actual matrix and to treat these solutions in an identical procedure as the unknown samples.

The calculations of the drug concentration of each analyte are performed identical to what has been shown in Box 9.1.

This example is based on Myasein F *et al.*[4].

Fundamental Questions to the Procedure—Step by Step

First try to answer the following questions by yourself, and subsequently read our discussion below:

1. Why is this both an identification method and a quantification method?
2. Why does the plasma contain K_2EDTA?
3. Why is the thawed sample mixed thoroughly?
4. Why are the samples stored at −80 °C?
5. Why is the sample preparation performed in 96-well plates?
6. Why is an internal standard added to all samples and calibration solutions?
7. Why is 100 μl of the mixture aspirated and dispensed five times?
8. Why is 200 μl acetonitrile used for protein precipitation?
9. Why is 100 μl of 3 M zinc sulfate also added?
10. Why is the 96-well plate centrifuged at 3000 g for 5 minutes at 10 °C?

11. Why is an organic layer formed?
12. Why is 100 µl of the organic layer mixed with 100 µl of water?
13. What is the chromatographic separation principle?
14. Why are 1.8 µm particles used?
15. Why is the flow rate 0.5 ml/min?
16. Why is MS/MS used for detection?
17. Why is the fragment of ritonavir and ritonavir-d_5 identical?
18. Why does the calibration range differ from analyte to analyte?
19. Why do the analytes show different LLOQs?

Discussion of the Questions

1. The purpose of this method is for monitoring clinical trials where quantitative data are needed. It is important that the quantitative data are reliable, and therefore it is important to make sure that the peak measured also has the correct identity.
2. K_2EDTA is present in the Vacutainer® in order to prevent the sampled blood from coagulation. EDTA binds the Ca^{2+} needed for coagulation, and therefore no coagulation takes place.
3. During freezing and thawing, some fractionation of the sample may take place. Therefore, it is important to mix the sample thoroughly to obtain a homogeneous sample.
4. Samples are normally stored at $-20\,°C$ unless a lower temperature ($-80\,°C$) is required. This is to improve sample stability. Often, samples are collected over a period of time and then analyzed. It is important that the analytes are stable during the storage period.
5. Sample preparation is performed in a 96-well plate format using multipipettes in order to increase throughput.
6. The internal standards are added in the same amount to the sample and calibration standards as early in the sample preparation procedure as possible in order to improve repeatability and accuracy. This is especially important when using detectors where the experimental conditions may vary over time.
7. Aspiration and dispensing are performed five times using a pipette in order to obtain a thorough mixing of sample and internal standard.
8. When protein precipitation is performed using acetonitrile, it is common to use one part of sample and two parts of solvent. This secures full precipitation of the proteins.
9. The addition of 3 M zinc sulfate serves two purposes. The zinc ions also precipitate proteins, and the resulting high ionic strength will make the mixture separate into two phases because the acetonitrile is no longer miscible with the water phase.
10. The mixture is centrifuged in order to remove the precipitated proteins and to obtain a full separation of the two liquid phases.

11. Normally acetonitrile and water are miscible, but when high salt concentrations are present in the water, the two solvents are no longer miscible.

12. The organic layer used contains a high concentration of acetonitrile. If this is injected directly into the LC system, it is much stronger eluting than the mobile phase (55% acetonitrile). In order to obtain an elution strength of the sample solution similar to the mobile phase in the LC system, the sample is mixed with an equal amount of water.

13. The chromatographic separation principle is standard reversed phase conditions.

14. The small particle diameter of the stationary phase provides higher separation efficiency per unit length of the column, and therefore a very short column can be used. This again will provide high-throughput analysis.

15. The flow rate of 0.5 ml/min is higher than the normal standard of 0.2–0.25 ml/min for 2.1 mm I.D. columns. However, the small particle diameter prevents any loss in efficiency due to the higher linear flow rate, and this will also add to the high-throughput analysis.

16. MS/MS is used to obtain as high selectivity as possible. The quantitative measurement of a fragment obtained from the initial isolated pseudo-molecular ion will eliminate most interference. This will improve the reliability of the data obtained.

17. The fragments used for the quantitative measurement of ritonavir and its internal standard are identical because they do not contain the molecular part where the deuteration was placed. Ritonavir and its internal standard can be measured separately because the parent ions have a difference of 5 Da.

18. The difference in calibration range for the two analytes is just a reflection of the wish to be able to deliver quantitative data at the lowest possible concentration. The calibration range is from the LLOQ, which differs for the two analytes.

19. A difference in LLOQ of the analytes is observed, and this is due to the fact that the two compounds have different response factors, which again is due to the difference in molecular structure.

Additional Comments to the Sample Procedure

Protein precipitation in plasma and serum samples with acetonitrile is a common technique and is often sufficient. But sometimes some additional treatment is used in order to further concentrate or purify the sample. In the present case, salting out is used to concentrate the analytes in the organic phase while the aqueous part can be discarded. Another possibility is to freeze the mixture at $-80\,°C$. This will also make it possible to isolate the acetonitrile phase while the aqueous phase is frozen. A third possibility is to add a small amount of a lipophilic solvent (e.g., chloroform, dichloroethane, or trichloroethane)—a small amount will make the phases separate, and a larger volume of the lipophilic solvent will extract all the acetonitrile and leave the aqueous phase similar in volume as the original sample but without the proteins. If the analysis is about very polar solutes, this is a way to remove the proteins without diluting the sample.

9.2   Whole Blood Samples

As discussed in this chapter, plasma and serum samples are very popular for quantification of small molecule drug substances. However, whole blood samples also are used, and these are particularly popular in the field of forensic toxicology. Whole blood samples cannot be injected directly into LC or gas chromatography (GC), as they will immediately clog and deteriorate the injection systems and the chromatographic columns. Thus, prior to the final instrumental analysis, whole blood samples have to undergo some type of sample preparation. Normally, the whole blood samples are first diluted with water or buffer, and then processed by LLE or SPE. Dilution is important in order to reduce the viscosity of the sample and to improve the compatibility with LLE or SPE. A practical example of a total procedure for analysis of small-molecule drugs from whole blood is described in Box 9.6. This is an example from a forensic toxicology laboratory.

Box 9.6 Screening for Drugs and Drugs of Abuse in Whole Blood Samples by SLE and UHPLC-MS/MS

Purpose and Procedure

The current procedure is intended for identification and quantification of 28 different drugs and drugs of abuse in whole blood samples. The method is important in order to uncover driving under the influence of drugs, clarify cause of death, or discover drug use (violence, rape, prison cases, and social cases). The procedure is based on supported liquid extraction (SLE) and UHPLC-MS/MS.

Target Analytes

Alprazolam, amphetamine, bromazepam, buprenorphine, carisoprodol, clonazepam, cocaine, codeine, diazepam, ethyl morphine, fenazepam, fentanyl, flunitrazepam, lorazepam, MDMA (3,4-methylenedioxymetamphetamine), meprobamate, methamphetamine, methadone, methylphenidate, midazolam, morphine, nordiazepam, nitrazepam, oxazepam, oxycodone, THC (Δ9-tetrahydrocannabinol), zolpidem, and zopiclone.

Chemical structure and physicochemical properties of some of the analytes:

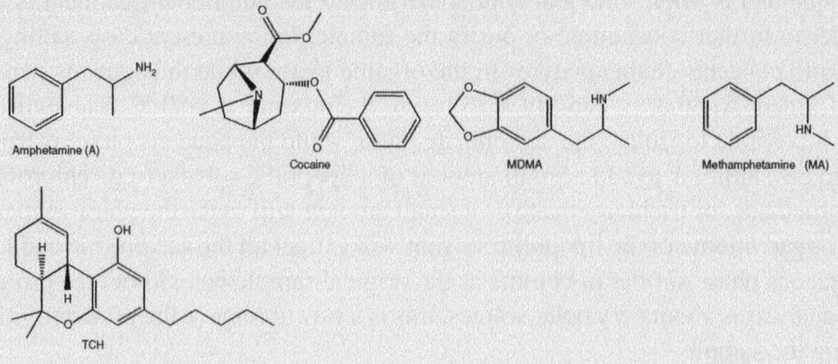

Amphetamine (A)   Cocaine   MDMA   Methamphetamine (MA)

TCH

	Amphetamine	Cocaine	MDMA	Methamphetamine	THC
log P	1.80	2.28	1.86	2.24	5.94
pK_a	10.01	8.85	10.14	10.21	9.34

Amphetamine, cocaine, MDMA, and methamphetamine are basic substances, whereas THC is slightly acidic.

Internal Standard Solution

Amphetamine-d_3 (511 ng/ml), buprenorphine-d_4 (42 ng/ml), clonazepam-d_4 (144 ng/ml), cocaine-d_3 (950 ng/ml), codeine-d_6 (366 ng/ml), MDMA-d_5 (357 ng/ml), methamphetamine-d_{11} (352 ng/ml), methadone-d_9 (350 ng/ml), morphine-d_3 (288 ng/ml), nordiazepam-d_5 (717 ng/ml), diazepam-d_5 (723 ng/ml), and THC-d_3 (41 ng/ml) in water.

Sample Collection and Pretreatment

Samples of whole blood are collected in 5 ml Vacutainer® tubes containing 20 mg sodium fluoride and sodium heparin. Autopsy samples are collected in 25 ml Sterilin tubes containing 200 mg potassium fluoride. The samples are stored at 4 °C prior to processing.

Sample Preparation

Five hundred microliters of whole blood are mixed with 50 µl internal standard solution, 250 µl 0.2 mol/l ammonium carbonate buffer, and 400 µl preservative (containing the surfactant Triton X-100). This sample solution is transferred to a column for SLE. The analytes are extracted from the column with two portions of ethyl acetate–heptane (4:1, v/v). The eluate is evaporated to dryness under N_2 at 60 °C. The residue is dissolved in 60 µl methanol–water (20:80, v/v).

UHPLC-MS/MS

UHPLC-MS/MS is performed under the following conditions:

Column	C18, 2.1 mm × 10 cm, 1.8 µm particles
Mobile phase	A: Methanol, B: 10 mM ammonium format in water (pH 3.1)
Flow rate	500 µl/min
Gradient	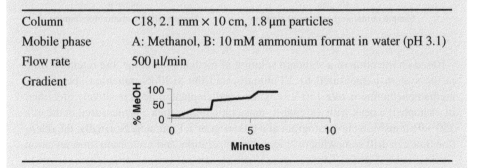

Run time	9 min	
Column temperature	65 °C	
MS polarity	Positive	
ESI	1 kV	
MS/MS transitions (selected analytes)	Amphetamine	136 → 91
	Cocaine	304 → 182
	MDMA	194 → 163
	Methamphetamine	150 → 91
	THC	315 → 193
Calibration range (selected analytes)	Amphetamine	4–400 ng/ml
	Cocaine	6–600 ng/ml
	MDMA	6–600 ng/ml
	Methamphetamine	4–400 ng/ml
	THC	0.3–13 ng/ml

Identification

Identification of drugs in blood samples is based on a combination of retention time data and MS/MS transition data. An example is illustrated here for identification of methamphetamine:

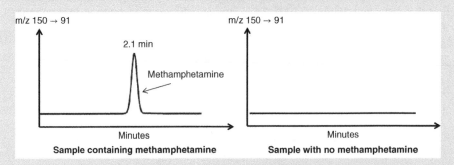

Based on injection of a standard solution of methamphetamine, the retention time in the system is measured to 2.1 minutes, and the MS/MS transition specific for methamphetamine is m/z 150→91. Thus, methamphetamine is positively identified in a sample if a peak with a signal-to-noise ratio higher than 3 is measured in the m/z 150→91 transition chromatogram at a retention of 2.1 minutes. Normally, the retention time can drift somewhat over long-term operation, and a retention time deviation within 1% can be accepted. The chromatogram to the left shows a sample containing methamphetamine, and the chromatogram to the right shows a sample with no methamphetamine.

Calibration and Quantitative Analysis

Quantification of drugs in blood samples is based on calibration curves. Each of the 28 target analytes has its own calibration curve. An example is illustrated for quantification of methamphetamine in the table and figure below. The calibration curve for methamphetamine is established in the concentration range of 4–400 ng/ml. Six drug-free blood samples are spiked with known and different concentrations of methamphetamine, and they are spiked with a known and constant concentration of the internal standard methamphetamine-d_{11} (calibration samples). The calibration samples are analyzed according to the method described above, methamphetamine (MA) is measured at the m/z 150→91 transition, and the internal standard (MA-d_{11}) is measured at the m/z 161→97 transition. The data are given below.

Calibration sample no.	MA (ng/ml)	MA-d_{11} ng added	150→91 peak area	161→97 peak area	Peak area ratio
1	4	17.6	8 094	72 883	0.11
2	25	17.6	51 006	71 092	0.72
3	50	17.6	103 760	73 394	1.41
4	100	17.6	210 584	74 126	2.84
5	250	17.6	526 467	75 481	6.97
6	400	17.6	854 353	72 664	11.76

The peak area ratio for the six calibration samples is plotted versus the concentration of methamphetamine to construct the calibration curve as shown below. The R^2 value is 0.999, and this indicates very good linearity. The concentration of methamphetamine (x) in collected blood samples analyzed in the same sequence as the calibration samples is calculated based on the following calibration equation based on the data above:

$$y = 0.0291x - 0.0588$$

where y is the measured peak area ratio. For a collected blood sample where the signal at m/z 150→91 is measured to 84 509 and the signal at m/z 161→97 is measured to 73 415, the concentration of methamphetamine is given by:

$$84\ 509/73\ 415 = 0.0291x - 0.0588$$

$$x = 41.6 \text{ ng/ml}$$

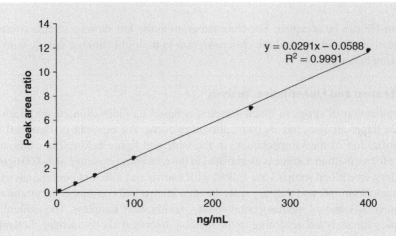

This example is based on Leere Øiestad *et al.* [5].

Fundamental Questions to the Procedure—Step by Step

First try to answer the following questions by yourself, and subsequently read our discussion below:

1. Why is this both an identification method and a quantification method?
2. What is a Vacutainer® tube?
3. Why does the Vacutainer® tube contain 20 mg sodium fluoride?
4. Why does the Vacutainer® tube contain sodium heparin?
5. What is a Sterilin tube?
6. Why does the Sterilin tube contain 200 mg potassium fluoride?
7. Why are samples stored at +4 °C?
8. Why is the sample volume 500 µl?
9. Why is internal standard added to the sample?
10. Why are deuterated internal standards used?
11. Why is ammonium carbonate buffer added to the sample?
12. Why is preservative added to the sample?
13. Why is supported liquid extraction used?
14. Why are ethyl acetate and n-heptane mixed?
15. Why is the eluate evaporated?
16. Why is the extract reconstituted in methanol–water?
17. Why is UHPLC used for separation?
18. Why is a C18-column used?
19. Why is the mobile phase composed of methanol and ammonium formate in water?
20. Why is gradient elution used?
21. Why is the column heated to 65 °C?
22. Why is MS/MS used for detection?

23. Why is the MS/MS operated in the positive ion mode?
24. Why is the ESI voltage set at 1 kV?
25. Why is methamphetamine measured at the m/z transition 150→91?
26. Why is methamphetamine-d_{11} measured at the m/z transition 161→97?
27. Why does the calibration range differ from drug to drug?

Discussion of the Questions

1. Compound identification is mandatory in forensic toxicology laboratories to identify drug abuse. Quantification in blood is important additional information as this gives an indication about the level of abuse.
2. Vacutainer® tube is a collection tube for whole blood. During sampling, a medical needle is connected to the tube, and vacuum inside the tube sucks whole blood through the needle and into the tube.
3. Sodium fluoride serves as a preservative, inhibits enzymes, and stabilizes target analytes in the whole blood sample.
4. Sodium heparin serves as an anticoagulant.
5. Sterilin tube is a sterile collection tube for biological samples. This tube is not based on vacuum. A Vacutainer® tube cannot be used for autopsy blood, due to the consistency of this material.
6. Potassium fluoride serves the same purpose as sodium fluoride in Vacutainer® (see point 3).
7. Storage in refrigerator improves the sample stability.
8. With blood samples, one tries to use as small volume as possible. In this method, 500 µl was found to be appropriate, and acceptable MS signals were obtained for all target analytes in their relevant concentration levels. With the newest generation mass spectrometers, even lower volumes of blood can be used due to improved sensitivity. Blood volumes less than 100 µl are less desirable due to difficulties of accurate pipetting.
9. The internal standards are added to the sample in order to improve repeatability and accuracy. This is especially important with MS instrumentation, where experimental conditions often vary over time.
10. Deuterated internal standards are perfect with LC-MS/MS because they have exactly the same physicochemical properties as the analytes, yet still they can be measured separately from the analytes in the MS instrument.
11. Ammonium carbonate buffer is added to the sample to make it slightly alkaline. This is important to suppress the ionization of the basic analytes. However, not all of the analytes are basic, and ammonium carbonate buffer was found to be a good compromise based on experimental optimization.
12. The preservative (surfactant) is added to improve the extraction of THC. This compound easily sticks to surfaces and can be difficult to analyze for this reason.
13. SLE is easier to automate, and the practical workflow is more straightforward as compared to LLE. Based on practical experience, SLE gives clearer extracts

from whole blood and autopsy blood than LLE, and this is another reason for the prevalence for SLE.

14. The target analytes are a mixture of weakly basic and acidic compounds. Therefore, the extraction solvent has to be a compromise. Mixtures of n-heptane and ethyl acetate are frequently used in forensic toxicology laboratories as a generic solvent mixture. Ethyl acetate gives appropriate polarity, and the solvent mixture is easy to evaporate.

15. The mixture of n-heptane and ethyl acetate cannot be injected in LC.

16. The mixture of methanol and water can be injected in LC. The use of methanol is important for the reconstitution of the more nonpolar analytes, as they are not dissolved in pure water.

17. UHPLC is used in order to get efficient separations under high mobile phase flow rates. Operation under high flow rates is beneficial because this reduces the analysis time per sample significantly.

18. Reversed phase chromatography is well suited for the target analytes, and for this separation principle C18-columns are very robust.

19. Methanol is used because it is less expensive and less toxic than acetonitrile. Ammonium formate is added to acidify the mobile phase and to stabilize pH. Acidic conditions in the mobile phase are important to charge the analytes prior to MS. Ammonium formate is compatible with LC-MS.

20. Gradient elution is used because the target analytes represent a large span in polarity, and therefore they cannot all be separated under isocratic conditions.

21. The column is heated to 65 °C to reduce the back pressure over the column. This is especially important in UHPLC, where separations are performed at high mobile phase flow rates.

22. MS/MS is used to obtain as high specificity as possible. This is mandatory in forensic toxicology.

23. The basic analytes form positive species in the mobile phase, and they are naturally detected as positive ions. Even THC, which is an acidic substance, can be detected in the positive mode.

24. The ESI voltage has been tested by experiments, and 1 kV was found to be appropriate.

25. The molecular weight of methamphetamine is 149 (M), and this substance is transferred as $(M + H)^+$ at m/z 150 through the first MS. In the second MS, fragmentation due to collision occurs, and a fragment at m/z 91 is formed according to the figure below. This fragment is transferred through the third MS and measured by the detector.

26. The internal standard is methamphetamine with 11 deuterium atoms. This transfers the first MS at m/z = 150 + 11 = 161. Six of the deuteriums are located

on the benzene ring fragment, and therefore m/z = 91 + 6 = 97 transfer the third MS.

27. Different drugs are consumed in different amounts based on their pharmacological and toxicological differences. Therefore, different drugs have different relevant concentration ranges.

9.3   Dried Blood Spots

Identification and quantification of drug substances in *dried blood spots* (DBS) are becoming increasingly popular, especially in the pharmaceutical industry. In the DBS approach, blood spot samples are collected by applying a few drops of whole blood, drawn by lancet from the finger, heel, or toe, onto a specially manufactured absorbent filter paper called a DBS card. DBS cards are typically cellulose-based materials, but also noncellulose materials have recently become commercially available. A typical DBS card is illustrated in Figure 9.3.

The blood is allowed to saturate the paper on the DBS card and is air-dried for a couple of hours. The DBS cards are stored in plastic bags with desiccant added to reduce humidity, and they may be kept at ambient temperature.

Once in the laboratory, the bioanalytical chemist punches a small disc of paper (saturated with dried blood) from the DBS card using an automated or manual punch. This disc is then subjected to sample preparation (desorption). The target analytes have to be desorbed from the dried blood spot and dissolved prior to the final analytical measurement (typically by LC-MS/MS). For this purpose, the punch is transferred to a desorption solvent. Desorption is a critical step, as (i) target analytes should be released from the dried matrix, (ii) target analytes should be completely dissolved in the desorption solvent, and (iii) preferably most matrix components from the blood sample should remain undissolved. Criteria (i) and (ii) are important in order to obtain high recovery, whereas criterion (iii) is important in order to obtain a reasonable degree of sample cleanup. For basic drugs, the desorption solvent is typically formic acid dissolved in a mixture of methanol and water. Desorption is followed by centrifugation to remove blood components and proteins released into the desorption solvent.

The exact blood volume loaded into the card is not very critical, because the blood sample is expected to distribute uniformly in the porous structure of the card. Exact blood volume

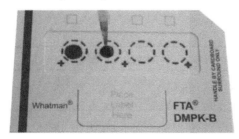

Figure 9.3   *Photo of dried blood spots (DBS) on specially manufactured absorbent filter paper (a DBS card).*

is achieved by punching the center of each spot with a 3 mm diameter puncher. This gives a uniform blood volume from spot to spot, and only a part of each spot is punched. The spots normally dry within a couple of hours, but overnight drying ensures totally dried samples.

DBS sampling has become very popular in recent years because of convenience. The DBS cards can be shipped easily in an envelope to a laboratory, which is much simpler than with traditional blood samples. Because the target analytes are stored in a dry state, the DBS cards can be stored and shipped under ambient temperature conditions. Until now, most DBS cards have been cellulose-based material (filter paper). However, very recently also noncellulose cards have been introduced. The advantages of noncellulose cards are (i) improved analyte recovery and (ii) better spot-to-spot homogeneity independently of the hematocrit value of the blood sample. The improved analyte recovery implies that target analytes are more easily desorbed from the DBS card. The better spot-to-spot homogeneity implies that blood samples with different hematocrit values, having different viscosity, are sucked into the porous DBS card material to the same degree. A practical example of a total procedure for analysis of small-molecule drugs from a DBS card is described in Box 9.7.

Box 9.7 Identification and Quantification of β-Blockers in Dried Blood Spots by LC-MS/MS

Purpose and Procedure

The current procedure is intended for quantification of atenolol, pindolol, metoprolol, and propranolol in DBS. The procedure is based on dried blood spot sampling and LC-MS/MS.

Chemical Structure and Physicochemical Properties of Some of the Analytes

	Atenolol	Pindolol	Metoprolol	Propranolol
log P	0.5	1.9	1.6	3.0
pK$_a$	9.6	8.8	9.7	9.5

All four target analytes are basic substances and are used as β-blockers.

Sampling

Fifteen microliters of whole blood samples are spotted on a dried matrix spotting (DMS) card (a non-cellulose-based DBS card). The spots are allowed to dry overnight. Circular punches of 3 mm are taken from the DBS cards for desorption.

Desorption

Each 3 mm punched DBS is desorbed using 300 μl of 0.1% formic acid in methanol and water (80:20 v/v). The sample is centrifuged at 15 000 rpm for 15 minutes. Subsequently, the supernatant is collected, evaporated, and reconstituted in 100 μl LC-MS/MS mobile phase.

LC-MS/MS

LC-MS/MS is performed under the following conditions:

Column	C18, 2.1 mm × 30 cm, 2.7 μm particles
Mobile phase	A: 0.1% formic acid in H_2O, B: MeOH
Flow rate	200 μl/min
Gradient	
Run time	3 min
Column temperature	Ambient
MS polarity	Positive
ESI	4 kV
MS/MS transitions	Atenolol $276.2 \rightarrow 145.1$
	Pindolol $249.2 \rightarrow 116.1$
	Metoprolol $268.2 \rightarrow 116.2$
	Propranolol $260.2 \rightarrow 116.1$

Identification, Calibration, and Quantitative Analysis

Identification, calibration, and quantification are performed the same way as discussed in Box 9.6.

This example is based on Arora *et al.*, "Improving Sensitivity of Basic Drugs in Dried Blood Spotting through Optimized Desorption," application note, Agilent Technologies.

Fundamental Questions to the Procedure—Step by Step

First try to answer the following questions by yourself, and subsequently read our discussion below:

1. Why is a noncellulose card used?
2. Why is the blood volume 15 µl?
3. Why is the card dried overnight?
4. Why is a 3 mm punch used?
5. Why is 300 µl of 0.1% formic acid in methanol and water (80:20 v/v) used as desorption liquid?
6. Why is the sample centrifuged at 15 000 rpm for 15 minutes?
7. Why is the desorption liquid evaporated and reconstituted in 100 µl LC-MS/MS mobile phase?
8. Why is LC used for separation?
9. Why is a C18-column used?
10. Why is the mobile phase composed of methanol and formic acid in water?
11. Why is gradient elution used?
12. Why is MS/MS used for detection?
13. Why is MS/MS operated in the positive ion mode?
14. Why is the ESI voltage set at 4 kV?

Discussion of the Questions

1. A noncellulose card is used rather than a cellulose-based card because desorption of the target analytes is more efficient from the former. Also, the volume of blood collected in a 3 mm punch is more homogeneous on a noncellulose card. The volume is less dependent on the viscosity and the hematocrit value of the blood sample.
2. Fifteen microliters of blood are sufficient to measure the target drugs at ng/ml when highly sensitive LC-MS/MS is used. Fifteen microliters of blood also give a reasonable size of the spot, which is easy to punch and work with from a practical point of view.
3. Overnight drying ensures that the blood spot is completely dry before transfer to the laboratory.
4. See the discussion under point 2.
5. This is a generic desorption solvent developed as a general solvent providing high and reproducible recoveries for many different drug substances. A high content of methanol is important in order to precipitate proteins from the blood sample, which are then removed by centrifugation.
6. The sample is centrifuged after desorption to remove blood cells and proteins. Centrifugation for 15 minutes at 15 000 rpm has been found appropriate based on experimental experience.

7. The concentration of methanol in the desorption solvent is 80%, and this gives poor chromatography when injected in LC-MS/MS. Therefore, the solvent is evaporated, and the sample is dissolved in LC-MS/MS mobile phase (20% methanol). Reconstitution in 100 µl gives a small enrichment relative to the analyte concentration in the desorption solvent.
8. Several of the target analytes are relatively polar and are most suited for separation by LC.
9. A C18-column is used to give as strong retention as possible based on hydrophobic interactions. This is especially important for atenolol, which is a relatively polar analyte.
10. A mixture of methanol and water is used as mobile phase because the chromatographic principle is reversed phase chromatography. Methanol is less expensive and toxic than acetonitrile. Formic acid is added to the mobile phase to control pH (acidify the mobile phase). Under acidic conditions, the target analytes are positively charged, and this is important in connection with the ESI interface.
11. Gradient elution is used for two reasons. First, the target analytes have very different log P-values, and complete separation under isocratic conditions is difficult. In addition, by finishing each LC run with 80% methanol in the mobile phase, the column is efficiently rinsed after every run to remove matrix components from the dried blood samples. The latter is important to maintain the separation efficiency of the system.
12. MS/MS is used for highly specific and sensitive detection.
13. The MS/MS is operated in the positive mode because the model analytes (basic substances) are present as positive ions in the acidic mobile phase.
14. Four kilovolts are used as a standard setting for the ESI voltage.

9.4   Urine Samples

Urine samples are often used in forensic toxicology laboratories and in doping analysis laboratories for identification of drugs and drugs of abuse. The major advantage of urine samples is that no invasive sampling is required. Urine samples can be injected directly in LC or LC-MS, but sample preparation is often used in spite of this. For LC and LC-MS, the major purpose of sample preparation is to avoid loading the systems with matrix components, which can sacrifice instrumental stability. With LC-MS, avoidance of ion suppression is also an important argument in favor of sample preparation. For GC and GC-MS analysis, sample preparation of urine is mandatory to transfer the analytes to an organic extract, which can be injected into the system. Both LLE and SPE can be used for urine samples. A practical example of a total procedure for analysis of small-molecule drugs from urine is described in Box 9.8. This is an example from a doping control laboratory, and it is based on LLE and GC-MS.

Box 9.8 Screening for Anabolic Steroids in Urine Samples by LLE and GC-MS

Purpose and Procedure

The current procedure is intended for the identification and quantification of 30 different anabolic steroids in urine samples. The method is important in order to uncover doping in sports. The procedure is based on LLE and GC-MS.

Target Analytes

19-Norandrosterone, 19-noretiocholanolone, mesterolone metabolite, epitrenbolone, furazabol metabolite, epimethendiol, calusterone, bolasterone, clostebol metabolite, beta-boldenone, fluoxymesterone metabolite, 17α-methyltestosterone metabolite 1, 17α-methyltestosterone metabolite 2, ethylestrenol, oxandrolone, epioxandrolone, mibolerone, norethandrolone metabolite, metenolone, metenolone metabolite, stanazolol metabolite, boldenone metabolite, danazol, oxabolone, 1-testosterone, 4-hydroxy-testosterone, chlormetandienone metabolite, and metandienone metabolite.

Chemical Structure and Physicochemical Properties of Some of the Analytes

	19-Norandrosterone	1-Testosterone	Mibolerone	Calusterone	Oxabolone
log P	3.47	3.41	3.63	3.93	2.55

The analytes are neutral substances with no pK_a value.

Internal Standard Solution

17α-Methyltestosterone (200 ng/ml).

Derivatizing Reagent

N-methyl-N-trimethylsilyltrifluoroacetamide (MSTFA), NH_4I, and dithioerythritol (DTE) (1000:2:4, v/w/w).

Sample Preparation

Three milliliters of urine are mixed with 1 ml phosphate buffer pH 7.4, 50 μl internal standard solution, and 50 μl of beta-glucuronidase from *Escherichia coli*. This solution is incubated at 50 °C for 1 hour to complete hydrolysis of the glucoronide metabolites. After hydrolysis, 1 ml of carbonate buffer (pH 9) is added, and this solution is subjected to LLE. A volume of 10 ml tert-butyl methyl ether is used as extraction solvent, and LLE is performed for 6 minutes by shaking. After LLE, centrifugation is performed, and the organic layer is collected and evaporated to dryness. The residue is derivatized at 70 °C for 15 minutes by adding 50 μl of derivatizing agent. An aliquot of 1 μl of this solution is injected in GC-MS.

GC-MS

Carrier gas	Helium
Flow rate	1 ml/min
Injection type	Split injection
Split ratio	1:10
Injection port temperature	280 °C
Column	17 m length, 0.20 mm I.D., 0.11 μm film thickness fused-silica capillary column, 100% methyl silicone
Temperature program	
Transfer line temperature	280 °C
Detector temperature	280 °C
Ionization mode	Electron ionization, 70 eV
MS polarity	Positive
SIM diagnostic ions (selected analytes)	19-Norandrosterone 405 + 420
	1-Testosterone 194 + 179
	Epioxandrolone 308 + 363
	Mibolerone 301 + 446
	Calusterone 445 + 315
	Oxabolone 491 + 506
	Internal standard 301 + 446

Limits of detection	19-Norandrosterone	1 ng/ml
(selected analytes)	1-Testosterone	10 ng/ml
	Epioxandrolone	10 ng/ml
	Mibolerone	10 ng/ml
	Calusterone	10 ng/ml
	Oxabolone	10 ng/ml

SIM, selected ion monitoring.

Identification

Identification of target steroids in urine samples is based on a combination of retention time data and diagnostic ions. An example is illustrated for identification of calusterone in the figure below:

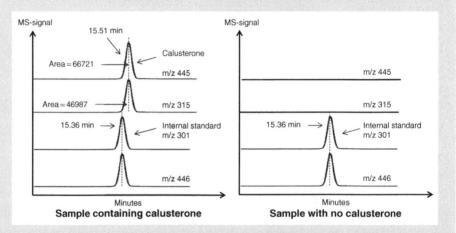

First, a standard solution (standard) is injected, and retention times for calusterone and internal standard (IS) are measured. In addition, peak areas for calusterone are measured at m/z 445 and 315, as illustrated below.

	Retention time			Peak area		
	Calusterone	IS	Ratio	m/z 445	m/z 315	Ratio
Standard	15.48	15.30	1.01	125 776	87 551	1.44
Sample	15.51	15.36	1.01	66 721	46 987	1.42

Based on injection of the standard solution, the relative retention time for calusterone is calculated to:

$$\text{Relative retention time} = 15.48/15.30 \text{ minutes} = 1.01$$

Based on injection of the standard solution, the peak area ratio (m/z 445/m/z 315) for calusterone is calculated to:

$$\text{Peak are ratio} = 125\ 776/87\ 551 = 1.44$$

Calusterone is positively identified in a sample if a peak with a signal-to-noise ratio higher than 3 is measured in the m/z 445 chromatogram and in the m/z 315 chromatogram, and with a relative retention time of 1.01 as compared with the internal standard. In addition, the peak area ratio for calusterone should be 1.44. Due to instrumental variations, up to a 1% deviation is accepted for the relative retention time, and up to a 10% deviation is accepted for the peak area ratio. The chromatograms to the left show a sample containing calusterone, and the chromatograms to the right show a sample with no calusterone. The peak area ratio for calusterone is 1.42, and this deviates $((1.42 - 1.44)/1.44) \cdot 100\% = -1.4\%$ from the standard.

Calibration and Quantitative Analysis

Calibration and quantification are performed the same way as discussed in Box 9.6.
This example is based on Mazzarino *et al.* [6].

Fundamental Questions to the Procedure—Step by Step

First try to answer the following questions by yourself, and subsequently read our discussion below:

1. Why is this both an identification method and a quantification method?
2. Why is the sample volume 3 ml?
3. Why is the sample mixed with phosphate buffer pH 7.4?
4. Why is internal standard added to the sample?
5. Why is 17α-methyltestosterone at 200 ng/ml used as internal standard?
6. Why is beta-glucuronidase added to the sample?
7. Why is hydrolysis performed at 50 °C and for 1 hour?
8. Why is 1 ml of carbonate buffer (pH 9) added to the sample after hydrolysis?
9. Why is LLE performed after hydrolysis?
10. Why is tert-butyl methyl ether used as solvent for LLE?
11. Why is the volume of tert-butyl methyl ether 10 ml?
12. Why is LLE performed for 6 minutes?
13. Why is centrifugation performed after LLE?
14. Why is the extract evaporated to dryness after LLE?
15. Why is the residue derivatized?
16. Why are MSTFA, NH$_4$I, and DTE used as derivatization reagents?
17. Why is derivatization performed at 70 °C and for 15 minutes?
18. Why is the injection volume in GC-MS 1 µl?
19. Why is GC-MS used?
20. Why is helium selected as carrier gas?
21. Why is the flow rate of carrier gas 1 ml/min?
22. Why is split injection used?

23. Why is the split ratio set to 1:10?
24. Why is the injection port temperature 280 °C?
25. Why is a 17 m length, 0.20 mm I.D., 0.11 μm film thickness fused-silica capillary column used?
26. Why is 100% methyl silicone used as the stationary phase?
27. Why is temperature programmed analysis used for the column?
28. Why is the transfer line temperature set at 280 °C?
29. Why is the detector temperature set at 280 °C?
30. Why is the mass spectrometer using electron ionization at 70 eV?
31. Why is the polarity of the mass spectrometer positive?
32. Why is the mass spectrometer operated in SIM mode?
33. Why are m/z 445 and 315 selected as diagnostic ions for calusterone?
34. Why is identification based on both retention times and peak area ratios?
35. Why are relative retention times used for identification?

Discussion of the Questions

1. Compound identification is mandatory in doping laboratories to identify a doping case. Quantification is important additional information as this can give some indication about the level of abuse.
2. In contrast to blood samples, urine samples can be collected in larger sample volumes. Thus, often several milliliters of urine are used, and the advantage of using such relatively large sample volumes is that analytes can be detected at low concentration levels.
3. Steroids are heavily bound to glucuronic acid when excreted in urine, and the first step in the method is an enzymatic hydrolysis where the steroids are released from glucuronic acid. For this enzymatic reaction to be efficient, pH has to be adjusted to pH 7.4.
4. Internal standard is added to the sample prior to sample preparation to improve the precision and the accuracy. The analytes are always measured relative to the internal standard.
5. 17α-Methyltestosterone is an appropriate internal standard because the structure is very close to the structure of the analytes of interest, and because this substance is not naturally found in urine. The concentration of the internal standard should be at the same concentration level as for the analytes, and therefore 200 ng/ml is selected.
6. Beta-glucuronidase is an enzyme that cleaves the binding of the steroids to glucuronic acid, and during the hydrolysis, the steroids are converted to their free form in the urine sample.
7. Heating to 50 °C for 1 hour is accomplished in order to make sure that the hydrolysis is completed to 100%.
8. pH is increased prior to LLE to improve the clean-up during LLE.
9. LLE is performed to isolate the steroids of interest from matrix components in urine and from the enzyme added to the sample. Most matrix components in

urine are highly polar and are not extracted into the solvent during LLE. This gives excellent sample clean-up.

10. Tert-butyl methyl ether is a commonly used solvent for generic methods in doping analysis. This solvent is easy to evaporate.

11. Ten ml of solvent is a relatively large volume for LLE of 5 ml aqueous phase, but this is selected to maximize extraction recoveries.

12. During method optimization, it has been found that 6 minutes is sufficient time to maximize extraction recoveries. The extraction is performed by strong shaking of the two phases.

13. Centrifugation is performed after LLE to facilitate a rapid separation of the two phases in the LLE system.

14. The 10 ml organic phase is evaporated, and the extract is dissolved in 50 μl derivatization reagent. This result in very efficient analyte enrichment and a more efficient derivatization reaction in a small volume.

15. The analytes have to be derivatized prior to GC analysis. Derivatization converts the analytes into less polar species and improves their gas chromatographic separation.

16. MSTFA, NH_4I, and DTE are a commonly used derivatization mixture for anabolic steroids. MSTFA is N-methyl-N-trimethylsilyltrifluoroacetamide, and the derivatization involves silylation of the free hydroxyl groups.

17. High temperature is used to increase the speed of the derivatization reaction. From experimental testing, it has been found that a 15-minute reaction time is sufficient at this elevated temperature to complete the reaction.

18. This is a typical injection volume in GC-MS. If larger volumes are injected, the system is easily overloaded, resulting in less efficient chromatographic separation.

19. GC is an excellent separation technique for derivatized steroids, because they are relatively nonpolar and volatile. MS is used for reliable identification.

20. Helium is selected as the carrier gas because this enables relatively high flow rates to be used without sacrificing separation efficiency. In addition, helium is a safe gas to handle in the laboratory.

21. With a column internal diameter of 0.20 mm (=0.02 cm), a 1 ml/min flow rate corresponds to:

$$1 \text{ ml/min} = 1 \text{ cm}^3/\text{min} = (1 \text{ cm}^3/\text{min})/(60 \text{ s/min}) = 0.017 \text{ cm}^3/\text{s}$$

$$= (0.017 \text{ cm}^3/\text{s})/(3.14 \cdot (0.02 \text{ cm}/2)^2) = 53 \text{ cm/s}$$

According to the van Deemter plot, this linear velocity is slightly higher than the optimal value for helium, and it has been selected to reduce the separation time.

22. Split injection is favorable when injecting dirty samples, because most of the sample matrix is effectively diverted to waste. Thus, split injection reduces the buildup of nonvolatile contamination of the injection port and the column.

23. The split ratio is 1:10. This means that 1 volume part of sample is diverted into the column, and 10 volume parts are diverted to waste. A relatively low split ratio

is used in this method in order to make sure that substantial amounts of sample enter the GC. This is important in order to detect low concentration levels of analyte.

24. The injection port temperature is 280 °C to make sure that sample constituents evaporate immediately during injection.

25. The column length is 17 m. This is a relatively short column and gives fast separation. The internal diameter of the column is 0.2 mm, and this is a relatively narrow column. A narrow column gives high chromatographic resolution and enables the system to be operated at high flow rates without sacrificing separation (see also point 21). The thickness of the stationary phase is 0.11 μm, and this is a standard dimension in GC.

26. The derivatized steroids are very nonpolar structures and can therefore be separated on a very nonpolar stationary phase.

27. Temperature programming of the column improves the separation of the different steroids. In addition, the column is heated to 320 °C during each analysis to make sure that all injected sample matrix is removed from the column prior to the next injection.

28. From the GC oven, the column is extended through a transfer line and into the mass spectrometer, which is more or less a separate unit (depending on the manufacturer). To make sure that analytes are not condensed in the transfer line (and do not reach the MS), this transfer line is heated to the same temperature as the injection port.

29. The detector temperature (inlet of the mass spectrometer) is also heated to 280 °C to make sure that analytes do not condense at the inlet of the system.

30. Standard GC-MS is performed with electron ionization, and the energy of the electrons responsible for ionization is adjusted to 70 eV.

31. The mass spectrometer is operated in the positive mode, because positive ions are formed during electron ionization.

32. The mass spectrometer is operated in the SIM mode, because the sensitivity is higher in SIM mode than in full-scan mode. The disadvantage with SIM mode is that full spectra are not recorded.

33. The molecular ion is at m/z 460 for derivatized calusterone, but m/z 445 and 315 are highly specific and much more abundant ions for this compound as compared to the molecular ion.

34. In all chromatography, retention times are used for identification. In this particular case, two characteristic ions are selectively measured for each steroid, and intensity ratios between two characteristic ions are as additional support for positive identification.

35. When retention times are measured relative to the retention time of the internal standard, they become more independent of small changes in the carrier gas flow rate and other experimental variations.

9.5   Saliva

Saliva samples are very attractive in the field of forensic toxicology. In contrast to blood samples, saliva samples can be collected in a simple, noninvasive manner by nonmedical personnel. Saliva samples can be collected under close supervision to prevent substitution or adulteration, which can be a problem with urine sampling. A practical example of a total procedure for analysis of small molecule drugs from saliva is described in Box 9.9. This is an example from a forensic toxicology laboratory.

Box 9.9 Screening for Drugs and Drugs of Abuse in Saliva Samples by LLE and LC-MS/MS

Purpose and Procedure

The current procedure is intended for identification and quantification of 32 different drugs and drugs of abuse in saliva samples. The procedure is based on LLE and LC-MS/MS.

Target Analytes

3-Hydroxy-diazepam, 6-monoacetylmorphine (6-MAM), 7-aminonitrazepam, 7-aminoflunitrazepam, alprazolam, amphetamine, benzoylecgonine, bromazepam, buprenorphine, carisoprodol, clonazepam, cocaine, codeine, diazepam, fenazepam, flunitrazepam, lorazepam, lysergic acid diethylamide (LSD), 3,4-methylenedioxyamphethamine (MDA), 3,4-methylenedioxyethylamphethamine (MDEA), MDMA, meprobamat, methamphetamine, methadone, morphine, N-desmethyldiazepam, nitrazepam, oxazepam, THC, zolpidem, and zopiclone.

Chemical Structure and Physicochemical Properties of Some of the Analytes

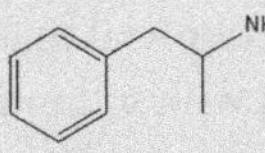

Amphetamine (A) 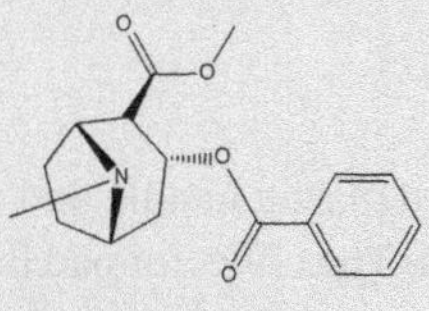   Cocaine 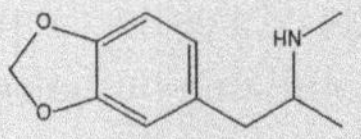   MDMA 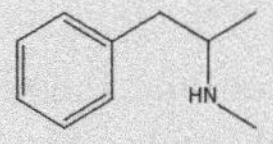   Methamphetamine (MA)

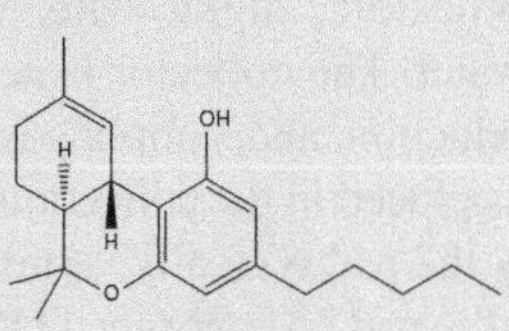

TCH

	Amphetamine	Cocaine	MDMA	Methamphetamine	THC
log P	1.80	2.28	1.86	2.24	5.94
pK$_a$	10.01	8.85	10.14	10.21	9.34

Amphetamine, cocaine, MDMA, and methamphetamine are basic substances, whereas THC is slightly acidic.

Internal Standard Solution

Morphine-d$_3$ (0.53 µmol/l), amphetamine-d$_3$ (3.0 µmol/l), methamphetamine-d$_{11}$ (0.85 µmol/l), benzoylecgonine-d$_8$ (0.31 µmol/l), MDMA-d$_5$ (0.77 µmol/l), 7-aminoflunitrazepam-d$_7$ (0.018 µmol/l), N-desmethyldiazepam-d$_5$ (0.045 µmol/l), and methadone-d$_9$ (0.27 µmol/l).

Sample Collection and Pretreatment

A special oral fluid collection kit is used as illustrated in the figure below:

The device consists of a collector pad on a plastic handle and a collection vial that contains 0.8 ml of stabilizing buffer solution. The collector pad is treated with sodium chloride, citric acid, sodium benzoate, potassium sorbate, gelatin, sodium hydroxide, and deionized water. The collection vial contains chlorhexidine digluconate, a blue dye, Tween 20 (nonionic surfactant), and deionized water. The collector plate is wiped between gum and cheek to stimulate the saliva production, and sampling is performed for 2 minutes. After sampling, the collector pad is placed in the collection vial for storage at +4 °C. The volume of saliva collected by the pad typically varies between 0.05 and 0.7 ml. The amount of saliva collected by the pad is determined by weighing the pad before and after sampling. The content of the pad is centrifuged, and 0.5 ml of the supernatant is transferred to a 5 ml polypropylene tube.

Sample Preparation

Five hundred microliters of saliva are mixed with 50 μl of internal standard solution and 250 μl of 0.2 mol/l ammonium carbonate buffer. The sample is extracted with 1.3 ml of ethylacetate/heptane (4:1) by mixing for 10 minutes. The mixture is then centrifuged at 1400 g for 5 minutes, and the organic phase is collected and evaporated to dryness under nitrogen at 40 °C. The residue is then dissolved in 60 μl acetonitrile–water (10:90, v/v) and injected into LC-MS/MS.

LC-MS/MS

LC-MS/MS is performed under the following conditions:

Column	C18, 2.1 mm × 5 cm, 3.5 μm particles	
Mobile phase	A: Acetonitrile, B: 5 mM ammonium acetate in water (pH 5)	
Flow rate	300 μl/min	
Gradient		
Run time	10 min	
Column temperature	35 °C	
MS polarity	Positive	
ESI	1 kV	
MS/MS transitions (selected analytes)	Amphetamine	135.9 → 91.0
	Cocaine	303.9 → 182.0
	MDMA	193.9 → 163.0
	Methamphetamine	149.9 → 91.0
	THC	315.1 → 193.1
Calibration range (selected analytes)	Amphetamine	0.5–5 μmol/l
	Cocaine	0.025–0.25 μmol/l
	MDMA	0.2–2 μmol/l
	Methamphetamine	0.2–2 μmol/l
	THC	0.005–0.05 μmol/l

Identification, Calibration, and Quantitative Analysis

Identification, calibration, and quantification are performed in the same way as discussed in Box 9.6.

This example is based on Leere Øiestad *et al.* [7].

Fundamental Questions to the Procedure—Step by Step

First try to answer the following questions by yourself, and subsequently read our discussion below:

1. Why is this both an identification method and a quantification method?
2. What is the purpose of citric acid and sodium benzoate in the collection pad?
3. What is the purpose of chlorhexidine digluconate and the blue dye in the collection vial?
4. Why are samples stored at +4 °C?
5. How can we calculate the volume of saliva?
6. Why is the sample volume 500 µl?
7. Why is internal standard added to the sample prior to LLE?
8. Why are deuterated internal standards used?
9. Why is ammonium carbonate buffer added to the sample prior to LLE?
10. Why are ethyl acetate and n-heptane mixed as extraction solvent?
11. Why is the volume of extraction solvent 1.3 ml?
12. Why is the extraction time 10 minutes?
13. Why is centrifugation at 1400 g for 5 minutes used?
14. Why is the extract evaporated to dryness and reconstituted in 60 µl acetonitrile–water (10:90, v/v)?
15. Why is evaporation performed under an atmosphere of nitrogen?
16. Why is HPLC used for the separation?
17. Why is a C18-column used?
18. Why is the LC column only 5 cm long?
19. Why is the LC column with 2.1 mm internal diameter?
20. Why is the mobile phase composed of acetonitrile and ammonium acetate in water?
21. Why is gradient elution used?
22. Why is MS/MS used for detection?
23. Why is the MS/MS operated in the positive ion mode?
24. Why is the ESI voltage set at 1 kV?
25. Why is methamphetamine measured at the m/z transition $150 \rightarrow 91$?
26. Why is methamphetamine-d_{11} measured at the m/z transition $161 \rightarrow 97$?
27. What does the 0.5–5 µmol/l calibration range for amphetamine correspond to in ng/ml?

Discussion of the Questions

1. Compound identification is mandatory in forensic toxicology laboratories to identify drug abuse. Quantification in saliva is important additional information as this gives an indication about the level of abuse.
2. Citric acid is used to stimulate the saliva production, and sodium benzoate is used as preservative.

3. Chlorhexidine digluconate is used to avoid bacterial growth, and the blue dye is added as a visual inspection of the authenticity of the sample (not exchanged with water).

4. The samples are stored at +4 °C to keep them cold and to avoid chemical degradation during storage.

5. The weight of the saliva sample is determined using an analytical balance, and this number is converted to volume assuming the same density as water.

6. Saliva samples are collected in relatively small volumes. Five hundred microliters was found to be an appropriate volume, and acceptable MS signals were obtained for all target analytes in their relevant concentration levels.

7. The internal standards are added to the sample in order to improve repeatability and accuracy. This is especially important with MS instrumentation, in which experimental conditions often vary over time.

8. Deuterated internal standards are perfect with LC-MS/MS because they have exactly the same physiochemical properties as the analytes, and still they can be measured separately from the analytes in the MS instrument.

9. Ammonium carbonate buffer is added to the sample to make it slightly alkaline. This is important to suppress the ionization of the basic analytes. However, not all of the analytes are basic, and ammonium carbonate buffer was found to be a good compromise based on experimental optimization.

10. The target analytes are a mixture of strongly basic compounds, weakly basic compounds, and acidic compounds. Therefore, the extraction solvent has to be a compromise. Mixtures of n-heptane and ethyl acetate are frequently used in forensic toxicology laboratories as a generic solvent mixture. Ethyl acetate gives appropriate polarity, and the solvent mixture is easy to evaporate.

11. The total volume of the aqueous phase is 800 µl. To ensure an efficient extraction, a somewhat larger volume is used for the organic phase. By experimental testing, 1.3 ml organic solvent has been found to be appropriate.

12. From experimental testing, extraction recoveries have been found not to increase after 10 minutes of LLE, and therefore 10 minutes has been selected as the extraction time.

13. Centrifugation is used to facilitate phase separation between the aqueous and organic phases. The time duration and the speed have been optimized during method development.

14. The mixture of n-heptane and ethyl acetate cannot be injected in LC. The extract is reconstituted in a mixture of acetonitrile and water, which can be injected in LC. The use of acetonitrile is important for the reconstitution of the more nonpolar analytes, as they are not dissolved in pure water.

15. Evaporation under an atmosphere of nitrogen is performed in order to avoid contact with oxygen, which potentially can cause oxidation of sample constituents.

16. HPLC is used for separation because most target analytes contain polar functionalities and therefore cannot be analyzed directly by GC.

17. Reversed phase chromatography is well suited for the target analytes, and for this separation principle C18-columns are very robust.
18. A relatively short column is used in order to keep separation times as short as possible. Thus, with short separation times, more samples can be analyzed per hour.
19. A column that has a smaller internal diameter than standard HPLC columns (with 4.6 mm I.D.) requires a lower flow of mobile phase, and this gives an important saving in the consumption of mobile phase (acetonitrile).
20. Acetonitrile is used as a common organic modifier in mobile phases for reversed phase HPLC. Ammonium acetate is added to acidify the mobile phase and to stabilize pH. Acidic conditions in the mobile phase are important to charge the analytes prior to MS. Ammonium acetate is compatible with LC-MS.
21. Gradient elution is used because the target analytes represent a large span in polarity, and therefore they cannot all be separated under isocratic conditions.
22. MS/MS is used to obtain as high specificity as possible. This is mandatory in forensic toxicology.
23. The basic analytes form positive species in the mobile phase and are naturally detected as positive ions. Even THC, which is an acidic substance, can be detected in the positive mode.
24. The ESI voltage has been tested by experiments, and 1 kV was found as a good compromise under method testing.
25. The molecular weight of methamphetamine is 149 (M), and this substance is transferred as $(M + H)^+$ at m/z 150 through the first MS. In the second MS, fragmentation due to collision occurs, and a fragment at m/z 91 is formed according to the figure below. This fragment is transferred through the third MS and measured by the detector.

26. The internal standard is methamphetamine with 11 deuterium atoms. This transfers the first MS at $m/z = 150 + 11 = 161$. Six of the deuteriums are located on the benzene ring fragment, and therefore $m/z = 91 + 6 = 97$ transfers the third MS.
27. The molecular weight of amphetamine is 135.2 g/mol. Thus, 0.5 μmol/l is corresponding to:

$$0.5 \, \text{μmol/l} = 5 \cdot 10^{-7} \text{mol/l} \sim 5 \cdot 10^{-7} \text{mol/l} \cdot 135.2 \, \text{g/mol}$$

$$= 6.76 \cdot 10^{-5} \text{g/l} = 6.76 \cdot 10^{-5} \text{mg/ml} = 68 \, \text{ng/ml}$$

Similarly, 5 μmol/l is equivalent to 676 ng/ml.

References

1. Shou, W.Z., Pelzer, M., Addison, T. *et al.* (2002) *J. Pharm. Biomed. Anal.*, **27**, 143–152.
2. Singhal, P., Yadav, M., Winter, S. *et al.* (2012) *J. Chromatogr. Sci.*, **50**, 839–848.
3. Kuhn, J. and Knabbe, C. (2013) *Talanta*, **110**, 71–80.
4. Myasein, F., Kim, E., Zhang, J. *et al.* (2009) *Anal. Chim. Acta*, **651**, 112–116.
5. Øiestad, L., Johansen, U., Oiestad, A.M. *et al.* (2011) *J. Anal. Toxicol.*, **35**, 280–293.
6. Mazzarino, M., de la Torre, X., Botrè, F. *et al.* (2007) *Rapid Commun. Mass Spectrom.*, **21**, 4117–4124.
7. Øiestad, L., Johansen, U. and Christophersen, A.S. (2007) *Clin. Chem.*, **53**, 300–309.

References

10

Analysis of Peptide and Protein Drugs in Biological Fluids

Leon Reubsaet and Trine Grønhaug Halvorsen

School of Pharmacy, University of Oslo, Norway

In the first part of this textbook (Chapters 1–8), the different techniques in use for bioanalysis have been presented, with special focus on the basic principles. All this you have to read, understand, and remember before you read Chapters 9 and 10. Chapter 10 is intended to give selected examples on how to use the different techniques with a focus on peptide and protein drugs. Polypeptide drugs are increasing in the pharmaceutical market, and Chapter 10 is included to give a taste of the use of chromatography within this field. The examples have been carefully selected to represent the varying complexity of analytes, ranging from small peptide drugs consisting of fewer than 20 amino acids to larger protein drugs consisting of more than 150 amino acids. As illustrated by the examples, different sample preparation strategies are applied depending on the complexity (size) of the polypeptide, the sample matrix, and the concentration level. Each example is provided with detailed explanations related to the experimental procedures. Hopefully, this will give the reader the understanding of all the small details that are difficult to get from more general textbooks in analytical chemistry. You should not try remembering all the details in the procedures from Chapters 9 and 10, but rather focus on the understanding. Bioanalytical methods vary a lot, but if you understand the examples given in Chapters 9 and 10, you have a very good platform for understanding most of the methods in use.

Box 10.1 describes analysis of a small cyclic peptide drug (11 amino acids) routinely monitored in whole blood due to its narrow therapeutic window. Due to the size of this peptide, conventional sample preparation techniques such as protein precipitation can be used.

Bioanalysis of Pharmaceuticals: Sample Preparation, Separation Techniques and Mass Spectrometry,
First Edition. Steen Honoré Hansen and Stig Pedersen-Bjergaard.
© 2015 John Wiley & Sons, Ltd. Published 2015 by John Wiley & Sons, Ltd.

Box 10.1 Quantification of Cyclosporine A in Whole Blood

Purpose and Procedure

The current procedure is intended for therapeutic drug monitoring of cyclosporine A, a cyclic peptide. Due to the narrow therapeutic window, the lack of correlation between dose and drug concentration, variable pharmacokinetics, and the large interaction potential, cyclosporine A is routinely monitored to maintain concentrations within the therapeutic range. The drug is extensively distributed in erythrocytes and therefore has to be quantified in whole blood. The procedure is based on hemolysis of blood cells followed by protein precipitation, injection of the supernatant, and determination by ultrahigh-performance liquid chromatography–mass spectrometry (UHPLC-MS)/mass spectrometry (MS).

Target Analytes

Cyclosporine A

Chemical Structure of the Analyte

Internal Standard Solution

Cyclosporine A-d4 (20 ng/ml in acetonitrile)

Sample Preparation

First, 100 µl of whole blood is transferred into a 1.5 ml test tube. Then, 2% zinc sulfate solution (150 µl) is added. The sample is vortexed for 10 seconds before the addition of 250 µl acetonitrile containing cyclosporine A-d4 (internal standard). The mixture is vortexed for 1 minute and centrifuged for 2 minutes at 14 000 rpm.

Subsequently, the supernatant is evaporated to dryness and reconstituted in 100 µl of mobile phase A. The supernatant is transferred into a liquid chromatography (LC) vial, and 15 µl are used for injection.

LC-MS/MS

LC-MS/MS is performed under the following conditions:

Column	C18, 2.1 mm × 50 mm, 1.7 µm particles
Temperature	80 °C
Mobile phase	A: 2 mM ammonium acetate with 0.1% formic acid (v/v) in H_2O; B: 2 mM ammonium acetate with 0.1% formic acid (v/v) in methanol
Flow rate	500 µl/min
Gradient	$T_{0.00-0.05}$ A: 90%; B: 10% $T_{0.50}$ A: 50%; B: 50% $T_{1.50-2.50}$ A: 1%; B: 99% $T_{2.85-3.00}$ A: 90%; B: 10%
Run time	3 min
MS polarity	Positive
ESI	2.5 kV
Mass analyzer	Triple quadrupole
MS/MS transitions	Cyclosporine A 1219.9 → 1202.7 (Quantifier transition) Cyclosporine A 1219.9 → 1184.8 (Qualifier transition) Cyclosporine A-d_4 1223.9 → 1206.5

ESI, Electrospray ionization; MS, mass spectrometry.

Understanding the Procedure

Cyclosporine A is a cyclic peptide consisting of 11 amino acids. The compound is neutral and rich in hydrophobic amino acids such as methylleucine. Due to extensive distribution of cyclosporine A into erythrocytes (40–60%), analysis is performed on whole blood. The first step of the procedure is therefore to hemolyze the blood cells to make the cyclosporine A available for analysis. Hemolysis is performed by adding 150 µl of a 2% zinc sulfate solution to the whole blood sample. The mixture is then vortexed for 10 seconds to ensure sufficient mixing. During hemolysis, the content of blood cell is released, making a correct determination of cyclosporine A possible. Subsequent zinc sulfate addition, acetonitrile (containing ISTD), is added to precipitate plasma proteins. A volume of 250 µl acetonitrile is used. This volume is within the normal volume range used for protein precipitation with acetonitrile (normal ratio of 1:1 to 1:2; here, sample volume (100 µl) + zinc sulfate solution (150 µl) = 250 µl – acetonitrile volume = 250 µl – ratio 1:1). The zinc sulfate added to hemolyze the blood cells will also assist the protein precipitation due to a salting-out effect. After addition of acetonitrile, the mixture is vortexed for 1 minute to ensure sufficient mixing, then subsequently centrifuged for 2 minutes at 14 000 rpm to settle the precipitate. Cyclosporine A is a small peptide and will not precipitate upon addition of acetonitrile. A discussion on using protein precipitation as a sample preparation technique for peptides can be found in Box 10.3. After protein precipitation and centrifugation, the whole supernatant is transferred to a new vial and evaporated to dryness. The residue is reconstituted in mobile phase A.

The analysis is performed using UHPLC (particle size, 1.7 µm) followed by MS/MS on a triple quadrupole. The LC is performed based on reversed phase chromatography using a C18 column, and using a mixture of ammonium acetate and formic acid in water and methanol as the mobile phase. The analysis is performed at 80 °C. The reason for this is a conformational change in cyclosporine A, resulting in peak broadening at lower temperatures. Cyclosporine A analysis is therefore normally performed at high temperature (75–80 °C) in order to achieve narrow peaks.

The mobile phase consists of formic acid and ammonium acetate in addition to water–methanol. The target analyte is a neutral cyclic peptide and will therefore not be protonated by such in the mobile phase. However, the presence of ammonium acetate promotes the formation of ammonium adducts. Immunosuppressants are often determined as adducts, and ammonium adducts are preferred over sodium adducts as they generally provide a better fragmentation pattern. The ammonium adducts formed result in a positive charge of the analyte prior to the MS, and the analytes are therefore detected as positive ions in positive MS polarity mode. The target analytes are detected by selected reaction monitoring (SRM), and two transitions are monitored for cyclosporine A. One transition is used for quantification purposes (quantifier transition), and the other transition is used as an additional confirmation of the presence of cyclosporine (qualifier transition). Adduct formation is described

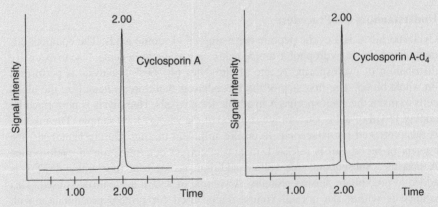

Figure 10.1 *Extracted ion chromatograms of a blood sample fortified with cyclosporine A (50 ng/ml) (chromatogram left) and added internal standard (chromatogram right). Reproduced with permission of Elsevier.*

to be less reproducible, but validation of the presented method shows acceptable repeatability (<10% relative standard deviation (RSD) at all levels).

An example of LC-MS chromatograms for the analytes (50 ng/ml fortified) in whole blood is illustrated in Figure 10.1.

This example is based on Tszyrsznic *et al.* [1].

Box 10.2 describes the analysis of several small peptide doping agents in urine. Due to the size of the peptides, conventional sample preparation techniques such as solid phase extraction (SPE) can be used. A high-resolution mass spectrometer in full-scan mode is used for detection, and by such the data can be evaluated retrospectively for the detection of new metabolites or other similar analytes.

Box 10.2 Quantification of Growth Hormone–Releasing Peptides in Urine

Purpose and Procedure

The current procedure is intended as a screening method in doping control of growth hormone–releasing peptides (GHRPs) and their major metabolites in human urine. The procedure is based on SPE and LC combined with high-resolution MS.

Target Analytes

Amino acid sequences of the GHRPs, their metabolites, and the ISTDs are shown here:

Ala-His-(d-β-Nal)-Ala-Trp-(d-Phe)-Lys-NH$_2$	GHRP-1
(d-Ala)-(d-β-Nal)-Ala-Trp-(d-Phe)-Lys-NH$_2$	GHRP-2
(d-Trp)-Ala-Trp-(d-Phe)-NH$_2$	GHRP-4
Tyr-(d-Trp)-Ala-Trp-(d-Phe)-NH$_2$	GHRP-5
His-(d-Trp)-Ala-Trp-(d-Phe)-Lys-NH$_2$	GHRP-6
Ala-His-(d-Mrp)-Ala-Trp-(d-Phe)-Lys-NH$_2$	Alexamorelin
His-(d-Mrp)-Ala-Trp-(d-Phe)-Lys-NH$_2$	Hexarelin
Aib-His-(d-2-Nal)-(d-Phe)-Lys-NH$_2$	Ipamorelin
(d-Ala)-(d-β-Nal)-Ala-NH$_2$	GHRP-2 metabolite
(d-3Ala)-(d-β-Nal)-Ala-NH$_2$	ISTD1
(d-Trp)-d4Ala-Trp-(d-Phe)-NH$_2$	ISTD2

GHRP, growth hormone–releasing peptide; ISTD, internal standard.

Sample Preparation

Sample preparation is performed using mixed-mode SPE with a combination of reversed phase and weak cation exchange. The pH of the urine samples is, if necessary, corrected to pH 7 by addition of acetic acid (2%) or sodium hydroxide (0.02 M) solution. Two milliliters of urine are fortified with 20 ng of ISTD1 and ISTD2 (20 µl of ISTD working solution). First, the SPE cartridge is conditioned with 1 ml of methanol and 1 ml of water. After sample addition, the cartridge is washed with 1 ml of water and 1 ml of methanol. Analytes are eluted using 1 ml of methanol containing 5% formic acid. The solvent is evaporated in a vacuum centrifuge (at approximately 40 °C), and the residue is resolved in 50 µl of 0.2% formic acid in water/acetonitrile (95%/5%). After centrifugation (17 000 g), the supernatant is transferred into an LC vial, and 10 µl are used for injection.

LC-MS

LC-MS is performed under the following conditions:

Column	C8, 1.0 mm × 50 mm, 1.9 µm particles	
Guard column	C8	
Mobile phase	A: 0.2% formic acid in H_2O	B: acetonitrile
Flow rate	120 µl/min	
Gradient	$T_{0.0}$ A: 95%; B: 5%	
	$T_{8.0}$ A: 65%; B: 35%	
	$T_{8.01-10.0}$ A: 80%; B: 20%	
	$T_{10.01-15.0}$ A: 95%; B: 5%	
Run time	15 min	
MS polarity	Positive	
ESI	Voltage not described	
Mass analyzer	Orbitrap	
MS-mode	Full scan ($m/z = 300 - 1000$, $R_s = 50\,000$)	
Mass range extracted ions	GHRP-1	478.25–478.26 (2+)
	GHRP-2	409.72–409.73 (2+)
	GHRP-4	608.28–608.29 (1+)
	GHRP-5	771.35–771.37 (1+)
	GHRP-6	437.22–437.23 (2+)
	Alexamorelin	479.75–479.76 (2+)
	Hexarelin	444.23–444.24 (2+)
	Ipamorelin	712.39–712.40 (1+)
	GHRP-2 metabolite	358.17–358.18 (1+)
	ISTD1	361.19–361.20 (1+)
	ISTD2	612.32–612.33 (1+)

ESI, Electrospray ionization; GHRP, growth hormone–releasing peptide; ISTD, internal standard; MS, mass spectrometry.

Understanding the Procedure

In this procedure, SPE by the mixed-mode principle is used as sample preparation in a screening method for several GHRPs in urine. The described method is intended for screening of non-endogenous GHRPs for doping purposes and for evaluation of GHRP metabolism. Urine is the most widely used matrix in doping analysis due to its relatively simple sampling method. However, urine as matrix is rather nonuniform, varying in salt concentration and pH, depending on for instance diet and fluid intake.

Prior to sample preparation, urine pH is measured and adjusted to pH 7 when necessary. In mixed-mode SPE, establishment of hydrophobic interactions is most important during sample application to the cartridges. Peptides are at their most hydrophobic state at the isoelectric point, that is, when the net charge is zero. This pH will vary from peptide to peptide depending on the amino acid composition. As the urine pH may vary from subject to subject, it is important to ensure uniform urine pH from sample run to sample run, and hence the pH of all samples is adjusted to 7 prior to sample application.

Mixed-mode sorbents are excellent for cleanup of zwitterionic compounds such as peptides. Retention on the mixed-mode SPE cartridges is based on both hydrophobic and ionic interactions. Due to this additional sample, cleanup is possible. The first washing step consists of 1 ml of water, ensuring the removal of water-soluble substances not retained by hydrophobic interactions. The second washing step consists of 1 ml of methanol and removes substances not retained by ionic interactions with the negatively charged cation exchanger group. In the elution step, 1 ml of a combination of methanol (breaking hydrophobic interactions) and 2% formic acid (protonation of the carboxylic acid on the sorbent) is used. The organic acidic eluate (strong eluting solvent) cannot be injected directly to LC-MS based on reversed phase chromatography. Therefore, the eluate is evaporated in a vacuum centrifuge, and the residue is resolved in a low volume (50 µl) of 0.2% formic acid in water/acetonitrile (95%/5%). Prior to injection, the resolved eluate is centrifuged to remove nonsoluble impurities after SPE. The large volume difference between sample volume (2 ml urine) and reconstitution solvent (50 µl) adapts for high preconcentration (maximum 40 times), and, taking the extraction recoveries into consideration, preconcentration ranging from 19 to 38 times is achieved.

After sample preparation, the samples are analyzed by LC followed by high-resolution MS. The LC is performed based on conventional reversed phase chromatography with a C8 column, using a mixture of formic acid in water and acetonitrile as the mobile phase. Because the target analytes are peptides with multiple ionizable groups, pH control in the mobile phase is important to maintain stable retention times, and therefore formic acid is added to the mobile phase. Formic acid also serves to ionize (protonate) the analytes prior to the MS, and the analytes are therefore detected as positive ions in positive MS polarity mode. The polarity of the peptides differs, and therefore the LC separation is carried out with gradient elution. Sub–2 μm particles also result in improved efficiency, allowing for equivalent resolution in less time (or faster analysis). However, higher flow rates will result in higher back pressure in the system, requiring an analytical system capable of operating at higher pressure. The target analytes are detected by ion extraction after full-scan MS using a high-resolution MS. This gives flexibility when looking for new metabolites in a nontargeted way. An example of LC-MS chromatograms for the analytes (0.5 ng/ml fortified) in urine is illustrated in Figure 10.2.

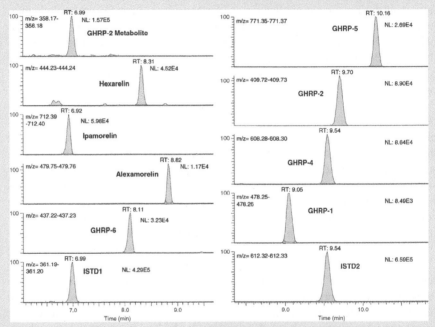

Figure 10.2 *Extracted ion chromatograms (width 0.01 Da) of a fortified urine sample (0.5 ng/mL) with abundant signals for (from top to bottom on the left) growth hormone–releasing peptide (GHRP-2) metabolite, hexarelin, ipamorelin, alexamorelin, GHRP-6, and internal standard 1 (ISTD1); and (from top to bottom on the right) GHRP-5, GHRP-2, GHRP-4, GHRP-1, and ISTD 2. Reproduced with kind permission from Springer Science and Business Media.*

This example is based on Thomas *et al.* [2].

Box 10.3 describes analysis of a polypeptide drug in plasma. Due to the size of the peptide (36 amino acids), conventional sample preparation techniques such as protein precipitation can be used. The method is intended for a clinical phase 2 study for establishment of the efficacy of the drug, and the common sample matrix for such experiments is human plasma.

Box 10.3 Quantification of Sifuvirtide in Human Plasma

Purpose and Procedure

The method described and explained is developed to determine the novel anti-HIV drug sifuvirtide in human plasma. This drug is a polypeptide. The intended goal was to quantify this biopharmaceutical in samples from a phase 2 clinical study. The method is based on protein precipitation followed by LC-MS/MS.

Target Analytes

The amino acid sequences of sifuvirtide and the internal standard are shown here:

Ac-SWETWEREIENYTRQIYRILEESQEQQDRNERDLLE-NH$_2$	Sifuvirtide
Ac-SWETWEREIENYTRQIYRILEENQEQQDRNERDLLE-NH$_2$	ISTD

ISTD, Internal standard.

Sample Preparation

A volume of 200 µl plasma is pipetted into a small vial. To this sample, 40 µl ISTD is added. After mixing, 0.4 ml acetonitrile is added. The vials are capped, mixed for 5 minutes, and centrifuged at 19 600 g for 10 minutes. The supernatant is transferred to a new tube and evaporated to dryness using nitrogen. The residue is reconstituted in 200 µl mobile phase A and, after mixing, again centrifuged at 19 600 g (5 minutes). A volume of 50 µl of this supernatant is injected into the LC-MS/MS.

LC-MS/MS

LC-MS/MS is performed under the following conditions:

Column	C18, 2.1 mm × 10 cm, 5 µm particles	
Guard	C18, 2.1 mm × 1.25 cm, 5 µm particles	
Mobile phase	A: 0.1% formic acid in H$_2$O; B: 0.1% formic acid in acetonitrile	
Flow rate	200 µl/min	
Gradient	T$_{0-1.0}$	A: 95%; B: 5%
	T$_{5.0}$	A: 5%; B: 95%
	T$_{5.5}$	A: 5%; B: 95%
	T$_{10}$	A: 95%; B: 5%
	T$_{12}$	A: 95%; B: 5%
MS polarity	Positive	
ESI	5 kV	
Mass analyzer	Triple quadrupole	
MS/MS transitions	Sifuvirtide	946.3 → 159.0
	ISTD	951.7 → 159.2

ESI, Electrospray ionization; ISTD, internal standard; MS, mass spectrometry.

Understanding the Procedure

Both sifuvirtide and its internal standard are chemically produced polypeptides. Their N-terminal serine (S) and C-terminal glutamic acid (E) are, respectively, acetylated and amidated forms.

The internal standard has a high degree of similarity compared to sifuvirtide. The only difference is the amino acid at position 23: For sifuvirtide this is serine (S), and for the internal standard this is asparagine (N). Both amino acid residues are both neutral and polar, implying that their charge and chromatographic behavior are similar. Sifuvirtide and its internal standard differ approximately 27 a.m.u. in mass.

Although it sounds contradictory, protein precipitation with subsequent supernatant analysis is chosen as the sample preparation. This relatively unspecific sample preparation can be used in this case because the expected concentration levels of sifuvirtide are in the medium ng/ml range and up. In analysis of small-molecule drugs, protein precipitation is used to get rid of all of the interfering proteins from a plasma or serum sample. Why are sifuvirtide and its internal standard not lost in the precipitate in this step? Both sifuvirtide and its internal standard are polypeptides consisting of 36 amino acids. Their solubility in organic solvents like acetonitrile is sufficiently high to keep them in solution during the acetonitrile-induced protein precipitation. This is tested experimentally. By adding 0.4 ml acetonitrile to 240 µl sample (200 µl plasma + 40 µl internal standard), the content of acetonitrile is approximately 62.5%. At this acetonitrile content, more than 99% of all plasma proteins will precipitate, leading to a good decrease in sample complexity. Because the sifuvirtide and its internal standard are still soluble at this high organic content, they will remain in the supernatant.

Because the supernatant has a high organic content, it is not possible to inject this into the LC-MS/MS system. The reason for this is that injecting a sample containing 62.5% acetonitrile in an LC gradient system, which starts with a mobile phase strength of 5% acetonitrile, will cause unnecessary band broadening. Therefore, the solvent is removed by evaporation. The nitrogen used for this purpose prevents oxidation of the sample components during this step. After the sample has dried, the residue is dissolved in LC-MS/MS mobile phase A to ensure compatibility. The centrifugation step after dissolving the residue is intended to remove particles and nondissolved components. This will prevent the LC-MS/MS system from being clogged.

The LC-MS/MS analysis consists of a straightforward separation of sifuvirtide and its internal standard from other interfering compounds on a reversed phase C18 column. Such interfering compounds might cause ion suppression or ion enhancement of the analyte and internal standard. The guard column is used to prevent the analytical column from being contaminated or damaged by the injected sample.

As mentioned in this chapter, both sifuvirtide and the internal standard are very similar in their amino acid composition and properties. Figure 10.3 shows chromatograms of both polypeptides. From this, it can be seen that both compounds elute approximately at the same time from the column.

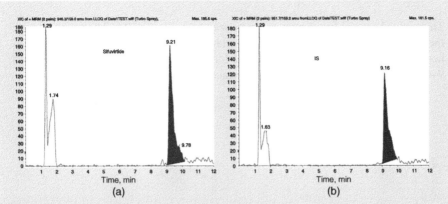

Figure 10.3 *Chromatograms of (a) sifuvirtide and (b) the internal standard. Reproduced with permission of Elsevier.*

Differentiation can be made by detecting them at different masses. Sifuvirtide has a mass of 4726.5, whereas the internal standard has a mass of 4753.5. Analyzing these polypeptides with electrospray will produce ions with multiple charges. This is visual in the mass spectrum as 5^+, 4^+, and, in the case of sifuvirtide, 3^+ ions. These multiple charges are due to the presence of several basic amino acid residues. Figure 10.4 shows the mass spectra after a full-scan analysis of both compounds.

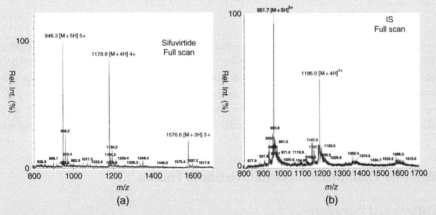

Figure 10.4 *Full-scan mass spectra of (a) sifuvirtide and (b) the internal standard. Reproduced with permission of Elsevier.*

In order to get high analytical selectivity, detection on the MS/MS was carried out in the SRM mode. The transitions were selected after a product ion scan of the fragmentation of sifuvirtide and the internal standard was performed. In both cases, the ion with the highest signal from Figure 10.4 ($[M+5H]^{5+}$) was chosen for fragmentation.

This yielded several fragments, as can be seen from Figure 10.5. The fragment with the highest intensity was chosen for quantitative purposes. For the method described here, this gave the following transitions: for sifuvirtide, mass-to-charge ratio (*m/z*) 946.3 → *m/z* 159.0; and, for the internal standard, *m/z* 951.7 → *m/z* 159.2.

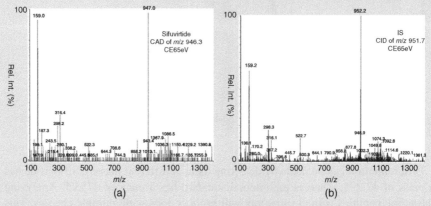

Figure 10.5 *Product ion scans of (a) sifuvirtide and (b) the internal standard after fragmentation. Reproduced with permission of Elsevier.*

This example is based on Che *et al.* [3].

Box 10.4 also describes the analysis of a polypeptide drug in plasma. The sample preparation of this peptide is a bit more complex than for the previous examples. After being subjected to SPE, the polypeptide is digested into smaller peptides prior to analysis by using a protease. One of the resulting peptides is then determined as a surrogate for the parent polypeptide. The digestion is performed to increase the sensitivity of the method, as the produced peptides normally have better ionization efficiency than their parent compound.

Box 10.4 Quantification of Enfuvirtide in Plasma

Purpose and Procedure

The method described and explained resembles the example in Box 10.3. The anti-HIV drug enfuvirtide is to be determined in human plasma. The drug is, as in the former example, a polypeptide. The method is based on SPE, proteolytic digestion, and LC-MS/MS.

Target Analytes

The amino acid sequences of enfuvirtide and the internal standard are shown here:

Ac-YTSLIHSLIEESQNQQEKNEQELLELDKWASLWNWF-NH$_2$ Enfuvirtide

Ac-YTSLIHSLIEESQNQQEKNEQELLELDKWASLWNWF-NH$_2$ Internal standard (d$_{60}$-enfuvirtide)

Sample Preparation

A volume of 500 µl plasma to which 10 µl internal standard is added is pipetted onto an activated Oasis HLB® SPE cartridge. After this, 1 ml of 0.1% trifluoroacetic acid is allowed to flow through, followed by 1 ml acetonitrile–water–trifluoroacetic acid (20:80:0.1). Elution is carried out using 1 ml acetonitrile–water–trifluoroacetic acid (60:40:0.1).

The eluate is evaporated to dryness using nitrogen, and the residue is reconstituted in 200 µl 100 mM Tris-HCl/2 mM CaCl$_2$ buffer (pH 8.6). To this, 5 µl 1.5 mg/ml chymotrypsin solution is added. The samples are subsequently incubated for 60 minutes at 37 °C. After this incubation, 4 µl 100% acetic acid is added. An aliquot of this sample is injected into the LC-MS/MS.

LC-MS/MS

LC-MS/MS is performed under the following conditions:

Column	C18, 2.1 mm × 5 cm, 3.5 µm particles	
Guard	C18, 2.1 mm × 1 cm	
Mobile phase	A: 0.25% formic acid in H$_2$O; B: 0.25% formic acid in methanol	
Flow rate	200 µl/min	
Gradient	T$_0$	A: 75%; B: 25%
	T$_6$	A: 60%; B: 40%
	T$_{8.5}$	A: 10%; B: 90%
	T$_{10.5}$	A: 10%; B: 90%
	T$_{10.51}$	A: 75%; B: 25%
	T$_{13}$	A: 75%; B: 25%
MS polarity	Positive	
ESI	5 kV	
Mass analyzer	Triple quadrupole	
MS/MS transitions	Enfuvirtide product A	525.2 → 376.1
	Enfuvirtide product C	476.2 → 118
	Enfuvirtide product E	804.1 → 1083.0
	Internal standard product A′	535.2 → 376.1
	Internal standard product C′	486.2 → 118
	Internal standard product E′	814.1 → 1088.5

ESI, Electrospray ionization; MS, mass spectrometry.

Understanding the Procedure

Both the enfuvirtide and its internal standard are chemically produced polypeptides. Their N-terminal tyrosine (Y) and C-terminal phenylalanine (F) are, respectively, acetylated and amidated forms. The internal standard (d_{60}-enfuvirtide) is in its amino acid composition identical to enfuvirtide; however, 60 deuterium atoms substitute 60 hydrogen atoms. This leads to a mass increase of approximately 60 compared to the nonlabeled enfuvirtide. Otherwise, charge and chromatographic behavior are similar.

Hydrophilic-lipophilic balance (HLB) was used for SPE material. This gave both nonpolar interactions as well as polar interactions with the polypeptides. As described in Section 6.5.3, the SPE material was first activated by washing with 1 ml methanol. Using 1 ml 0.1% trifluoroacetic acid, excess methanol is removed from the bed volume such that there will be good interaction between the applied sample and the SPE material. After sample application, all polar contaminants from plasma were removed by washing with 1 ml 0.1% trifluoroacetic acid. A next washing step using 1 ml acetonitrile–water–trifluoroacetic acid (20:80:0.1) allowed removal of compounds that had some hydrophobic interaction with the SPE material. Adding 1 ml acetonitrile–water–trifluoroacetic acid (60:40:0.1) allows enfuvirtide (and its internal standard) to elute. The moderate elution strength (60% acetonitrile) of the eluent accomplished efficient analyte removal, while many hydrophobic components remained on the SPE cartridge.

The sample was evaporated to dryness and reconstituted in 100 mM Tris-HCl/2 mM $CaCl_2$ buffer (pH 8.6). The buffer pH is optimal for proteolytic digestion with chymotrypsin. $CaCl_2$ was added to this buffer because it improves the activity of the proteolytic enzyme chymotrypsin. After chymotrypsin is added, enfuvirtide and the internal standard are transferred into smaller peptides. Chymotrypsin is used less frequently than the more common protease trypsin, but, due to its different selectivity, its use can sometimes be advantageous. Chymotrypsin cuts peptide bonds at the C-terminal of hydrophobic amino acids like leucine, isoleucine, phenylalanine, tyrosine, and tryptophan. In the sequence shown here, these amino acids are highlighted:

Ac-YTSLIHSLIEESQNQQEKNEQELLELDKWASLWNWF-NH$_2$

Although there are many possible cleavage sites, only a few products were monitored experimentally:

Product A	Ac-YTSL	(mass 525)
Product C	ASLW	(mass 475)
Product E	IHSLIEESQNQQEKNEQELL	(mass 2410)

The enzymatic reaction was stopped by changing the pH (by the addition of acetic acid) after only 1 hour of incubation. Incubation with proteolytic enzymes might be as long as overnight (18–20 hours), so the incubation time chosen in this example might be too short to allow complete cleavage of enfuvirtide. The yield of chymotryptic

products A, C, and E is, in this example, constant. This allows for quantitative determination of enfuvirtide by means of one (or more) of the chymotryptic products: the
production of products A, C, and E is stoichiometric to endufirtide.

Why is enfuvirtide digested into lesser peptides before analysis? This polypeptide might as well be analyzed directly after SPE, as is the case with sifuvirtide (see
Box 10.3). Often, peptide products after a proteolytic digestion have better ionization
efficiency than their parent compound. This allows lower detection and quantification
limits.

The LC-MS/MS analysis is based on reversed phase high-performance liquid
chromatography (HPLC). Chromatography on the C18 column will result in more
retention of the nonpolar peptides compared to the polar peptides. With the triple
quadrupole as mass spectrometric detector operating in the full-scan mode, peaks
can be assigned to the chymotryptic products (see Figure 10.6).

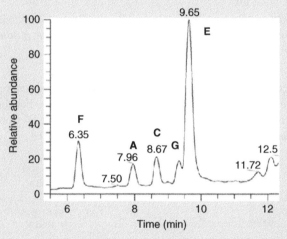

Figure 10.6 *Chromatogram of a digested enfuvirtide standard. Peak A, C, and E represent the measured product peptides A, C, and E described in this chapter. Reproduced
with permission of Elsevier.*

Quantification is performed by operating the triple quadrupole in the MS/MS mode.
Via a product ion scan analysis of the chymotryptic products of enfuvirtide and
the internal standard, the best mass transitions for SRM analysis were determined.
Figure 10.7 shows the product ion scan of the chymotryptic product E (enfuvirtide) as well as the chymotryptic product E' (internal standard). By fragmenting
m/z 804.1 ($[M+3H]^{3+}$ of enfuvirtide), it can be seen that ion m/z 1083.0 gives the
highest signal intensity. Thus, the transition for this chymotryptic peptide of enfuvirtide is 804.1 → 1083.0. This is similar for the internal standard: by fragmenting
m/z 804.1 ($[M+3H]^{3+}$ of enfuvirtide), it can be seen that ion m/z 1083.0 gives the
highest signal intensity. Thus, the transition for this chymotryptic peptide of enfuvirtide is 814.1 → 1083.0. The same procedure is carried out for the other chymotryptic

peptides of enfuvirtide and the internal standard, resulting in the mass transitions as described in the LC-MS/MS method.

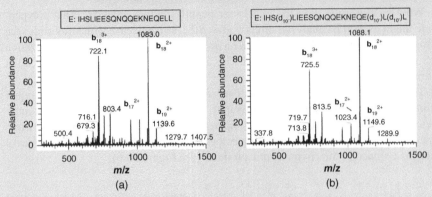

Figure 10.7 *Product ion scan of fragments resulting from (a) mass-to-charge ratio (m/z) 804.1 (chymotryptic product E) and (b) m/z 814.1 (chymotryptic product E'). Reproduced with permission of Elsevier.*

Subsequent LC-MS/MS analysis using the mass transitions discussed here allows determining the quantity of enfuvirtide using peak intensities. For the chymotryptic product E from enfuvirtide and the chymotryptic product E' from the internal standard, this results in chromatograms, as shown in Figure 10.8.

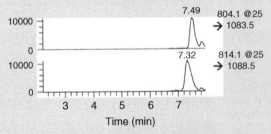

Figure 10.8 *Selected reaction monitoring (SRM) chromatograms of enfuvirtide (by means of the chymotryptic product E—upper trace) and the internal standard (by means of the chymotryptic product E'—lower trace). The difference in retention time between both peptides is caused by the deuterium effect (explained in Section 7.9.7). Reproduced with permission of Elsevier.*

This example is based on van den Broek *et al.* [4].

Box 10.5, the fifth and last example, describes the analysis of two protein drugs in plasma. Due to the size of the proteins (166 amino acids each) and the relatively low levels present in plasma, specialized sample preparation must be performed to separate the proteins of interest from the endogenous proteins present in plasma. This last example demonstrates how complex a sample preparation procedure can become for the determination of proteins in a protein-rich matrix such as plasma. It also illustrates some of the challenges that can occur during development of the method.

Box 10.5 Identification of Darbepoietin and Erythropoietin in Plasma

Purpose and Procedure

The method described here is a confirmative LC-MS/MS analysis of darbepoietin (DPO) and erythropoietin (EPO) in plasma. These are protein drugs used in the treatment of anemia as they stimulate the production of red blood cells. Because these drugs are abused in sports, methodology that can confirm their presence is required for doping analysis purposes.

Target Analytes

The amino acid sequences of both DPO and EPO are shown here:

DPO:
APPRLICDSRVLERYLLEAKEAENITTGCNETCSLNENITVPDTKVNFY
AWKRMEVGQQAVEVWQGLALLSEAVLRGQALLVNSSQVNETLQLH
VDKAVSGLRSLTTLLRALGAQKEAISPPDAASAAPLRTITADTFRKLF
RVYSNFLRGKLKLYTGEACRTGDR.
EPO:
APPRLICDSRVLERYLLEAKEAENITTGCAEHCSLNENITVPDTKVNFY
AWKRMEVGQQAVEVWQGLALLSEAVLRGQALLVNSSQPWEPLQLH
VDKAVSGLRSLTTLLRALGAQKEAISPPDAASAAPLRTITADTFRKLF
RVYSNFLRGKLKLYTGEACRTGDR.

Sample Preparation (1)

(a) Four hundred microliters of magnetic beads coated with anti-rhEPO antibody are added to 2 ml plasma. This is incubated for 16 hours at 37 °C under gentle shaking. Using a magnet, the magnetic beads are separated and the supernatant is discarded. After that, the beads are washed three times with phosphate-buffered saline (PBS) pH 7.4 containing the surfactant Igepal CA-630®.

(b) After this, 1 ml of 0.1% PEG6000 in PBS pH 2 is added to the magnetic beads. Using a magnet, the magnetic beads are separated from the supernatant. The supernatant is collected and filtered over a 0.22 µm membrane (this under centrifugation at 2500 g for 5 minutes).

(c) The filtrate is transferred to a filter device with a molecular-weight cut-off of 30 kDa. This device is centrifuged at 3500 g for 35 minutes. The filtrate is discarded. To the same device, 0.4 ml ammonium bicarbonate buffer (50 mM, pH 7.8) is added; the device is subsequently centrifuged at 3500 g for 20 minutes, and the filtrate is discarded. This is repeated four times. The remaining solution on top of the filter is collected for digestion.

Sample Preparation (2)

(d) To 80 μl of extract in ammonium bicarbonate buffer, 10 μl trypsin (200 μg/ml) is added. This mixture is vortexed and placed at 37 °C for 3 hours. The mixture is then placed at 80 °C for 10 minutes.

(e) In the last step, 5 μl PNGase F (20 000 U/ml) is added and incubated at 37 °C for 1 hour. After this incubation, 4 μl formic acid (10%) is added, and after that 20 μl acetonitrile.

An aliquot of this sample is injected into the LC-MS/MS.

LC-MS/MS

LC-MS/MS is performed under the following conditions:

Column	C18, 1 mm × 5 cm, 3.53 μm particles, 300 Å
Guard	C18, 1 mm × 1.7 cm, 5 μm particles
Mobile phase	A: 0.1% formic acid in water/acetonitrile (95/5)
	B: 0.1% formic acid in water/acetonitrile (5/95)
Flow rate	50 μl/min
Gradient	T_0 A: 100%; B: 0%
	T_1 A: 100%; B: 0%
	T_{11} A: 80%; B: 20%
	T_{20} A: 71%; B: 29%
	$T_{20.5}$ A: 20%; B: 80%
	T_{26} A: 20%; B: 80%
	$T_{26.5}$ A: 100%; B: 0%
	T_{32} A: 100%; B: 0%
MS polarity	Positive
ESI	5 kV
Mass analyzer	Triple quadrupole
MS conditions	Full scan m/z range 300–1500
MS/MS conditions	DPO 765.7 → 600–1200
	EPO 787.4 → 600–1200

DPO, darbepoietin; EPO, erythropoietin; ESI, electrospray ionization; MS, mass spectrometry; m/z, mass-to-charge ratio.

Understanding the Procedure LC-MS and LC-MS/MS Analyses

The LC-MS/MS analysis is based on reversed phase HPLC. Chromatography on the C18 column will result in more retention of the nonpolar peptides compared to the polar peptides. A gradient was used to be able to determine both the hydrophilic peptides as well as the hydrophobic peptides in one single analysis. With the triple quadrupole as mass spectrometric detector operated in the full-scan mode (m/z range,

300–1500), peaks can be assigned to the tryptic products. (Both Figures 10.9 and 10.10 are results of this single MS analysis. These figures will be discussed later.)

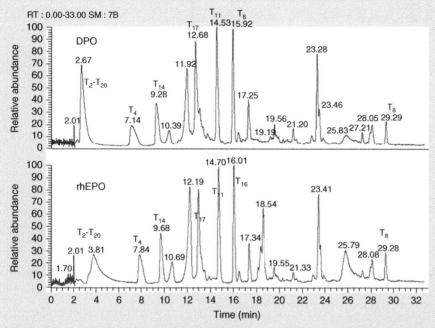

Figure 10.9 *Chromatograms of digested (a) darbepoietin (DPO) and (b) erythropoi-etin (EPO). Reproduced with permission of Georg Thieme Verlag KG.*

When the triple quadrupole mass spectrometer was operated in the MS/MS mode, only the *m/z* values belonging to DPO tryptic peptide number 9 (T9DPO) and EPO tryptic peptide number 9 (T9EPO) were selected (765.7 and 787.4, respectively) and subsequently fragmented. By scanning the products of this fragmentation in the range of 600–1200, specific fragments for these peptides were recorded (see Figure 10.11; this figure will be discussed later).

Sample Preparation (1)

(a) Magnetic beads coated with anti-rhEPO allow affinity extraction of both DPO and EPO because the antibody has affinity for both proteins: due to the affin-ity, these proteins bind strongly (but not covalently) to the antibodies, which again are covalently bound to the magnetic beads. The optimal temperature often is 37 °C. A simple magnet is then used to retain the magnetic beads (which contain the extracted DPO or EPO). After this step, interfering proteins and compounds are washed away using buffer (pH 7.4) but also a mild detergent (Igepal CA-630®).

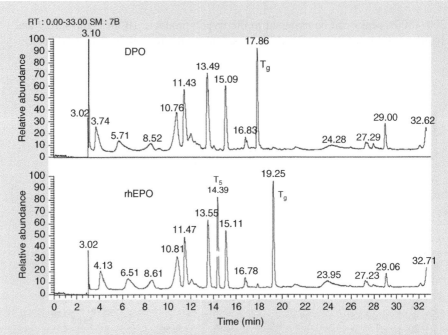

Figure 10.10 *Chromatograms of digested and deglycosylated darbepoietin (DPO; upper chromatogram) and erythropoietin (EPO; lower chromatogram). Reproduced with permission of Georg Thieme Verlag KG.*

(b) To release the extracted proteins from the antibodies, the pH is lowered to pH 2. This causes both conformational changes as well as charge changes in the antibody and DPO/EPO, leading to interaction loss. In other words, the protein is released from the antibody. Using the magnet, the magnetic beads are separated from the supernatant, which in this case contains the released DPO/EPO. To ensure that remaining beads in the supernatant will not interfere with the separating system, the supernatant is pressed through a 0.22 μm filter.

(c) The filtrate from step (b) is transferred to a 30 kDa filter and centrifuged. In this way, the proteins will remain on top of the filter while the pH 2 buffer will go through it. To repeat the process with 50 mM ammonium bicarbonate pH 7.8 is to exchange the pH 2 solution with a buffer suitable for trypsination.

Sample Preparation (2)

There are two enzymes used to allow differentiation between DPO and EPO: trypsin and PNGase F. To understand the choice and mechanism of these enzymes, one should take a look at the amino acid sequence of DPO and EPO first. These look identical; however, there are some small but distinct differences. There are five places that are different, and these are underlined in the amino acid sequence. When choosing the bottom-up approach using trypsin, the amino acid sequence will be cut on the C-terminal of arginine and lysine (**R** and **K**, respectively). As described under

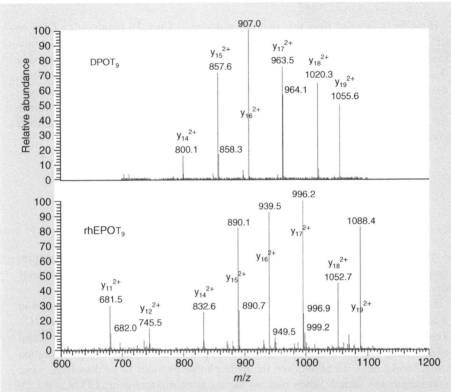

Figure 10.11 *Mass spectrometry (MS)/MS spectra of deglycosylated T9 from (a) darbepoietin (DPO) and (b) erythropoietin (EPO). Reproduced with permission of Georg Thieme Verlag KG.*

the procedure, the proteolytic enzyme trypsin is added to protein at a pH of around 8. This pH is the optimal one for trypsin. DPO will yield 21 different peptides. The same with EPO. To make it easier, the peptides are numbered as follows: T1 for APPR (i.e., amino acids 1–4 in the sequence), T2 for LICDSR (i.e., amino acids 5–10 in the sequence), and so on until T21 (TGDR). Many of the peptides that are generated will be identical; thus, it will be impossible to discriminate between if they originate from DPO or from EPO: T1 from DPO has the same amino acid sequence as T1 from EPO. More specific, only T5 and T9 show differences.

T5$_{DPO}$	EAENITTGCNETCSLNENITVPDTK
T5$_{EPO}$	EAENITTGCAEHCSLNENITVPDTK
T9$_{DPO}$	GQALLVNSSQVNETLQLHVDK
T9$_{EPO}$	GQALLVNSSQPWEPLQLHVDK

After performing the tryptic digest and the subsequent LC-MS analysis, many of the products are visible (see Figure 10.9).

However, it is clear from this figure that neither T5 nor T9 (from either DPO or EPO) is detectable. These are the peptides, which are essential to be able to discriminate between DPO and EPO. The reason why T5 and T9 are not being detected is that these peptides are glycosylated. This means that, due to biochemical processes, a sugar chain has been coupled to one or more amino acids. This results in glycopeptides with a different retention time and m/z ratio compared to the expected ones for the bare peptide (without the sugar chains). For both T5 and T9, the (possible) glycosylation sites are shown here in bold:

$T5_{DPO}$	EAENITTGCNETCSLNENITVPDTK
$T5_{EPO}$	EAENITTGCAEHCSLNENITVPDTK
$T9_{DPO}$	GQALLVNSSQVNETLQLHVDK
$T9_{EPO}$	GQALLVNSSQPWEPLQLHVDK

An extra step to remove the sugar chains is therefore incorporated in the sample pretreatment. The enzyme, PNGase F, is added. It specifically cuts off the sugar chains from asparagine (N) simultaneously, converting asparagine to aspartic acid.

In the procedure described here, both DPO and EPO are digested using trypsin in a first step. The activity of trypsin is terminated by increased temperature (80 °C). This usually has little effect on the peptides itself. After this step, PNGase F is added to allow deglycosylation. This reaction is stopped by acidifying the digest solution using formic acid. To allow even the hydrophobic peptides to be dissolved, some acetonitrile was added. When running the resulting mixture using LC-MS, the T9 peptide suddenly is clearly visible (see Figure 10.10) in both the DPO and the EPO sample, whereas only T5 is visible in the EPO sample.

The difference in retention time between the T9 generated from DPO and the T9 peptide generated from EPO is caused by the difference in amino acid sequence. In this way, it is possible to discriminate between DPO and EPO.

To confirm the identity of $T9_{DPO}$ and $T9_{EPO}$, MS/MS was carried out. The fragmentation pattern is specific for the peptide and can be used to deduce the amino acid sequence. A description of the typical peptide fragmentation patterns can be found in Box 7.3 in Section 7.9.5.

Figure 10.11 shows the results of these MS/MS runs.

This example is based on Guan *et al.* [5].

References

1. Tszyrsznic W Borowiec A, Pawlowska E, et al., *J. Chromatogr. B* **928** (2013) 9–15
2. Thomas A, Höppner S, Geyer H et al., *Anal. Bioanal. Chem.* **401** (2011) 507–516
3. Che J Meng Q, Chen Z, et al., *J. Pharm. Biomed. Anal.*, **51** (2010) 927–933
4. van den Broek, I., Sparidans, R.W., Schellens, J.H. and Beijnen, J.H. (2007) *J. Chromatogr. B*, **854**, 245–259.
5. Guan F, Uboh CE, Soma LR et al., *Int. J. Sports Med.*, **30** (2009) 80–86

11

Regulated Bioanalysis
and Guidelines

Martin Jørgensen[1] and Morten A. Kall[2]

[1]*Drug ADME Research, H. Lundbeck AS, Valby, Denmark*
[2]*Department of Bioanalysis, H. Lundbeck AS, Valby, Denmark*

This chapter discusses regulated bioanalysis and guidelines in detail. The basic terms of validation have already been discussed in Chapter 5, and therefore this chapter is an extension of it. This chapter discusses the historical evolution of regulated bioanalysis, bioanalytical method validation (BMV), pre-study validation, in-study validation, documentation, regulatory requirements, and quality systems in regulated bioanalysis. The chapter has been written with the pharmaceutical industry's perspective.

11.1 Introduction

The process of developing a new drug from *drug discovery* to *drug development* typically covers a period of more than 10 years. During this period, bioanalysis is an important tool, as discussed in Chapter 1. In the discovery phase, bioanalysis is operated on a generic, nonregulated platform that supports rapid, internal decision making by providing investigative data (e.g., on metabolic stability) and pharmacokinetic (PK) data from animal in vivo pharmacological models. As the development of a new drug enters preclinical development, bioanalysis becomes focused on producing systemic exposure data that will be associated with data from the nonclinical safety studies. Nonclinical studies may include investigation on single- and repeated-dose toxicity, reproduction function, embryo/fetal and perinatal toxicity, and mutagenic and carcinogenic potential of the drug. Bioanalysis in these studies are conducted according to Good Laboratory Practice (GLP).

In early clinical development and first dose in man (*first human dose* or *first in man*) studies, the bioanalytical data from nonclinical safety studies serve as a safety liaison

Bioanalysis of Pharmaceuticals: Sample Preparation, Separation Techniques and Mass Spectrometry,
First Edition. Steen Honoré Hansen and Stig Pedersen-Bjergaard.
© 2015 John Wiley & Sons, Ltd. Published 2015 by John Wiley & Sons, Ltd.

to the safety of human volunteers. Hence, systemic exposure to the drug in test animals dosed with the highest dose that does not result in systematic adverse events (NOAEL = *no observed adverse effect level*) provides guidance for dose escalation in man and is the basis of calculating an exposure ratio, which is the ratio between the exposure of test animals in toxicity studies and the human exposure of the drug.

During clinical development, metabolites formed from the drug by the human are identified, and for some metabolites it may be relevant to demonstrate that the human exposure is covered by the toxicity test-animals. This procedure is denoted *metabolites in safety testing* (MIST) and is covered by dedicated regulatory guidance.

In a number of clinical studies, the pharmacokinetics of the drug and metabolites are the primary objectives of the study. These studies are, among others, drug–drug interaction studies, in which the impact of co-administered drugs is investigated regarding the pharmacokinetics of the drug, or vice versa; and ultimately the bioequivalence studies (BEs), in which the rate and extent of absorption of, for example, two formulations of the same drug are compared.

Bioanalytical support to studies that are part of a drug application file, conducted by guidance from dedicated regulatory bioanalytical guidelines and often conducted under GLP or Good Clinical Practice (GCP), is referred to as *regulated bioanalysis*. However, the quality systems GLP and GCP do not provide guidance on the requirements to the performance of the bioanalytical methods; this area is strictly regulated and covered by dedicated regulatory bioanalytical guidelines. The science around the bioanalytical methods and regulatory requirements has evolved during the last 25 years, and today (2014) it is an extensive regulated area.

11.2   The Evolution of Regulated Bioanalysis

In 1990, the first BMV workshop was arranged by the American Association of Pharmaceutical Scientist (AAPS), the US Food and Drug Administration (FDA), and a number of other organizations. The purpose of the workshop was to establish a dialog between the pharmaceutical industry and regulatory agencies, dedicated to investigating and harmonizing the procedures required in regulated bioanalysis.

Before 1990, there was no guidance and little consensus on how to validate bioanalytical methods, conduct bioanalysis, or present data to regulatory agencies. The *workshop* defined *accuracy, precision, selectivity, sensitivity, reproducibility, limit of quantification,* and *stability* as essential parameters for bioanalyical method validation. One of the most important outcomes of the first workshop was definitions of acceptance criteria for specific validation parameters.

The workshop conclusions were published in several scientific journals; however, the workshop report was not an official document of the FDA, and therefore the FDA decided to develop and publish draft guidance in 1998, to ensure awareness of the importance of BMV.

The second BMV workshop arranged by AAPS and FDA was held in 2000, and it also was called the Crystal City meeting due to its location in Washington, DC. This workshop focused on the experiences with BMV and the advances in analytical technology that had occurred in the 10-year period since the first BMV workshop.

Bioanalysis was becoming *hyphenated*, for example liquid chromatography–tandem mass spectrometry (LC-MS/MS), with the advantages of increased sensitivity and selectivity. With the introduction of LC-MS/MS, new challenges entered the field of BMV. Thus, incomplete clearance of the collision cell for ions between data points, as internal standards (ISs) and related metabolites often have the same product ion as the analyte ion, resulted in *cross-talk* between channel monitoring and reduced the data quality. Also, isobaric interferences from compounds with similar fragmentation patterns (metabolites, endogenous compounds, and concomitant medication) caused cross-talk, and interference from matrix components causing ion suppression (also termed matrix effects) comprised new terms or dimensions of bioanalysis that led to comprehensive scientific work and discussions for the following 5–10 years. The workshop introduced different categories of validation—*partial validation, cross-validation*, and *full validation*—and introduced *ligand binding assays* as a bioanalytical tool.

The Crystal City meeting in 2000 formed the basis for the *FDA Guidance for Industry: Bioanalytical Method Validation*, published in May 2001. The following AAPS/FDA Crystal City meeting in 2006 resulted in a white paper published in 2007 by Viswanathans *et al.* ("Quantitative Bioanalytical Methods Validation and Implementation: Best Practices for Chromatographic and Ligand Binding Assays") that has served as a de facto guidance to the industry. The white paper provided clarification on a large number of issues (e.g., stability, metabolites, matrix suppression, and reporting) and discussed implementation of a test for matrix factor (MF) that would quantify the magnitude of matrix effect in LC-MS/MS, in contrast to the 2001 guidance supporting a more qualitative test of interferences or matrix effect. Furthermore, the requirements to investigate reproducibility in samples from dosed subjects (*incurred samples*) were introduced.

At the AAPS/FDA meeting in February 2008, *incurred sample reanalysis* (ISR) was implemented as an in-study validation and a random process check of method performance. This resulted in comprehensive activities in the bioanalytical societies and reflections in the literature, because no specific guidance on the conduct and acceptance of the reanalysis was provided. Furthermore, this would also cause 5–10% additional analysis for the bioanalytical laboratory. A number of publications have suggested procedures for ISR, and today there is a consensus in the bioanalytical society that ISR is a valuable tool for monitoring the performance of the method. ISR has added to the quality of bioanalytical data.

A decade after the FDA issued their first guidance, the European Medicines Agency (EMA) followed with their guideline for BMV in 2011. The guideline is, in principle, based on the FDA 2001 guidance and has adopted the issues discussed at the AAPS/FDA Crystal City meetings. The ISR and MF were adapted in the EMA guideline, and acceptance criteria in line with current practice in the bioanalytical community were formulated.

A few inventions were introduced with the EMA guideline, among these the legal aspect to require a claim of GLP in BMV studies and to investigate any MF in hemolyzed and hypolipedemic plasma. Details on a number of tests have been added in the EMA guideline, and the requirements for a number of parameters are more specific in the EMA guideline compared to the FDA 2001 guidance, such as a target on the magnitude of carry-over and the number of different matrix sources to be investigated in validation.

A FDA-issued draft on an updated BMV guidance in September 2013 has received comprehensive feedback from the bioanalytical society in connection to the Crystal City meeting in December 2013. At the time of writing this chapter, the EMA guideline should be considered as the lighthouse in terms of global bioanalytical compliance. The guidance has, besides in Europe, been adapted in Canada. Japan and China may issue their own guidelines, but they will most likely align with the EMA guideline. The Brazilian authorities, ANVISA (Agência Nacional de Vigilância Sanitária), issued resolution No. 27 in 2012 as guidance for BMV. The ANVISA guidance is in line with the EMA guideline in many aspects, but it differs slightly in terms of MF experiments, and the ISR is not mentioned in the document.

11.3   Bioanalytical Method Validation

The performance of the analytical method employed for the quantitative determination of drugs and their metabolites in biological samples plays a significant role in evaluation and interpretation of the main endpoint of the sample analysis, namely, the pharmacokinetic data. Therefore, it is essential to employ well-characterized and fully validated analytical methods to provide reliable and unambiguous results. A method validation should be conducted after the final method optimization, as a systematic collection of the documentation that the performance characteristics of the method are suitable and reliable for the intended analytical applications, and can meet predefined acceptance criteria for each critical parameter.

A *full validation* is an establishment of all validation parameters to apply to sample analysis for the bioanalytical method for each analyte. However, in situations where minor changes have been applied to an already validated method, a full validation may not be necessary but could be replaced by a *partial validation*. A partial validation could range from as little as the determination of the within-run precision and accuracy to an almost full validation, and it could cover changed calibration concentration range, limited sample volume, change in matrix or species, change in anticoagulant, sample processing procedure, storage conditions, and so on. Changes in equipment or transfer of the bioanalytical method to another laboratory should, when conducted during an ongoing study, in addition involve elements of a *cross-validation* in order to demonstrate accuracy between methods, technologies, and laboratories. Furthermore, cross-validation should be considered when data within a study are obtained from different methods or from the same method operated in different laboratories.

The acceptability of analytical data during application of the method should correspond directly to the criteria used to validate the method. The validation approach is based on the philosophy that essential elements of a method performance validated prior to study conduct (*pre-study validation*) should be evaluated on a (analytical) daily basis during application of the method. Such in-study validation was formulated by the FDA in the draft Guidance for Industry from December 1998 together with the presentation of the basic parameters for a validation, including (i) accuracy, (ii) precision, (iii) selectivity, (iv) sensitivity, (v) reproducibility, and (vi) stability.

The pre-study validation should be conducted in order to demonstrate that a method aimed to be used for quantitative measurement of analytes (a specific chemical moiety being measured, i.e., the unchanged drug or a metabolite) in a given biological matrix (plasma, serum, etc.) is reliable and reproducible.

The terminology applied for BMV may differ from those used for method validation in other analytical disciplines. The guidance describes the different elements and sub-elements as these require a number of experiments to be conducted, including guidance on acceptance criteria and a pragmatic statistically evaluation. Some elements in the validation have changed during the years as technology has evolved, removing some analytical challenges and at the same time introducing new analytical uncertainties and challenges. One example is the use of assay specificity, which was introduced in 1990 as the ability of a method to measure only what it is intended to measure. At that time, fluorescence and UV detection were common detection techniques in HPLC (*high-performance liquid chromatography*), and the interferences from matrix, chemicals, and other drugs were potentially serious issues that had to be investigated and eliminated prior to validation. As LC-MS/MS was introduced, the selectivity of the mass detector was much higher than that of the photometric detectors and visible interactions became less dominant. However, a new and invisible interference became a significant issue in LC-MS/MS bioanalysis. The competition between matrix components (primary phospholipids) and the analyte in the ESI (electrospray ionization) source of the mass spectrometer was shown to cause suppression of the ionization of the analytes, denoted as *ion suppression*. The implications of ion suppression could affect the trueness of a reported analytical result without obvious signs, as standards and quality controls (QCs) based on control matrix would pass predefined acceptance criteria. The introduction of stable isotope-labeled internal standards was shown to be a main tool to eliminate the impact of ion suppression based on the assumption that the suppression on analyte and internal standard would covariate, and thus the internal standard would adjust for unknown and random variations in ionization. The specificity is no longer a part of the BMV vocabulary for LC-MS/MS analysis, but it has been replaced by selectivity as a measure of the variation of *quantitative matrix effect* (ion suppression or ion boost) and formulated as an internal standard normalized MF in the latest EMA guidance from 2011.

11.4   Pre-study Validation

The basic parameters for a BMV have not changed during the last three decades and are still defined as *accuracy, precision, selectivity, sensitivity, reproducibility,* and *stability*. These parameters were discussed briefly in Chapter 5, and in somewhat more detail in this section.

The *accuracy of a method* is defined as the degree of closeness of the determined value to the nominal or known true value under prescribed conditions. This is sometimes termed *trueness*. The accuracy should be evaluated by repeated analysis of four concentration levels covered by the method calibration line (the LLOQ (lower limit of quantification), low, medium, and high) analyzed in at least five determinations and replicated in three independent analytical runs. An analytical run can be defined as a complete set of study samples with an appropriate number of calibration standards and QC samples for their validation. Several runs may be completed in one day, or one run may take several days to complete. The quantification of the samples from the three analytical runs should be conducted by three independent calibration curves.

The *precision* is the closeness of agreement (degree of scatter) between a series of measurements obtained from multiple sampling of the same homogeneous sample under the same conditions as described for accuracy, and thus the accuracy and precision of the method should be based on the $4 \times 5 \times 3$ matrix. The accuracy and precision should be evaluated as inter- and intra-batch accuracy and variation.

The *selectivity* is defined as the ability of the bioanalytical method to measure and differentiate the analytes in the presence of components that may be expected to be present in the sample. These could include metabolites, impurities, degradation products, or matrix components. The method selectivity should be evaluated by two means: the presence of interfering peaks where the analyte is expected to appear in the chromatogram (in blank matrix), and the variation of the internal standard.

Whereas the selectivity describes the method's ability to distinguish between interferences from compounds and the analyte response, the sensitivity describes the ability to selectively distinguish between noise and the analyte.

The *sensitivity* is traditionally defined as the ability of a method to distinguish different concentrations. For methods with a linear response function, the sensitivity therefore is equal to the slope of the calibration curve. However, in regulated bioanalysis, *sensitivity* is defined as the lower limit of quantification. The LLOQ is the lowest amount of an analyte in a sample that can be quantitatively determined with a predefined and suitable accuracy and precision. This is a quantitative measure and should not be mixed with the limit of detection (LOD), which is defined as the lowest concentration of an analyte that can be detected by the method, but not necessarily be quantified.

The *repeatability* of a method is the precision of the method under the same operating conditions over a short period of time. The repeatability is often denoted as the *auto-sampler repeatability* or *auto-sampler stability*.

The *reproducibility* of a method is the precision of the method under the same operating conditions applied at two or more laboratories.

The *stability* is a measure of the analyte stability in the given sample, and is divided into short-term and long-term frozen stability.

The short-term stability is further divided into three elements: benchtop stability, freeze–thaw stability, and auto-sampler stability. The *benchtop stability* covers the duration and conditions under which the matrix samples are kept during sample preparation in the laboratory. This period of time usually corresponds to 12–24 hours. The *freeze–thaw stability* is investigated to mimic the scenario where the sample is thawed and frozen several times, for example in the case of reanalysis; usually, experiments are conducted to mimic at least three freeze–thaw cycles. The *auto-sampler stability* (postpreparative stability or auto-sampler reproducibility) is investigated to demonstrate that it is valid to re-inject the whole batch after one to three days in the auto-sampler to mimic the scenario where an analytical batch started on a Friday evening is stopped during the weekend due to instrumental malfunction, and is re-analyzed the following Monday.

Long-term frozen stability is conducted to demonstrate that the analyte(s) are stable during frozen storage at a given temperature (e.g., $-20\,°C/-80\,°C$) for a period that covers the expected storage period from when the sample is drawn until the final analysis. Stability in matrix may be dependent on the type of compound, matrix, temperature, container material, and so on, and it is not accepted to extrapolate between conditions. Thus, the stability study should mimic the exact conditions for the study samples.

The *calibration range* or *dynamic range* of a bioanalytical method is defined as the range from the LLOQ to the *upper level of quantification* (ULOQ) where a relationship between instrument response and known concentrations of the analyte should provide a linear relation that can be described by a simple algorithm. The calibration line should consist of at least six concentration levels, excluding blank and zero samples. The actual concentration of each calibration standard calculated from the resulting calibration curve (back-calculated value) should be compared to the nominal (or target) concentration of each calibrator. The difference between the actual and nominal concentration for each calibrant should meet predefined acceptance criteria for the accuracy.

Carry-over should be evaluated during the method validation to demonstrate that the method and instrumentation should be able to handle two consecutive injected samples, first the highest standard (ULOQ) followed by blank matrix samples, without this contributing to more than 20% of the peak area of the lowest calibration standard (LLOQ). The method may have a linear dynamic range of three decades and meet the criteria described under the calibration range. However, if the above requirements for carry-over cannot be met, then the dynamic range is too wide and the method should be modified. Thus, if the carry-over cannot be reduced, then the dynamic range should be reduced.

The *extraction recovery* (extraction efficiency) of an analytical process is reported as the percentage of the known amount of an analyte carried through the sample extraction and processing steps of the method. If the sample preparation step of the method utilizes a phase shift (solid phase extraction, or liquid–liquid extraction), then it is recommended to demonstrate during validation that the extraction recovery is consistent. If the method sample preparation consists of a protein precipitation followed by centrifugation and injection into the LC system, then it can be argued that the recovery is 100%. There is no clear guidance to when or how to decide if extraction recovery should be validated. The different elements of a method validation are summarized schematically in Table 11.1.

Additional elements in the method validation have been added during the evolution of regulated bioanalysis. The most important elements are ISR, carry-over, dilution integrity, MF, and selectivity toward hemolyzed and hypolipidemic matrix.

Matrix effects should be investigated and quantified as IS-normalized MF by calculating the ratio of the peak area in the presence of matrix (measured by analyzing blank matrix spiked after extraction with analyte) to the peak area in the absence of matrix (pure solution of the analyte). The IS-normalized MF also should be calculated by dividing the MF of the analyte by the MF of the IS. The experiment should be conducted at two concentrations corresponding to a maximum of three times the LLOQ and close to the ULOQ, and it should be conducted on six different lots of matrix. In addition to the normal matrix, it is recommended to investigate matrix effects on other samples, for example hemolyzed and hyperlipidemic plasma samples.

Dilution integrity is to demonstrate that an attempted dilution is conducted accurately and precisely. Dilution may involve several steps and involve pipettes and robots that have to be calibrated toward the applied volume, actual liquid density, and so on. Dilution integrity experiments should be investigated in the pre-study validation as well as in the in-study process, as this is a dynamic process.

Table 11.1   Elements of a Method Validation

Validation parameter	Pre-study experiment	Acceptance criteria	In-study experiment	Acceptance criteria	Introduced
Calibration line	A minimum of six calibration concentration levels exclusive blank and a zero sample	At least 75% of the calibration standards should be ±15% of the nominal value, except for the LLOQ for which it should be within ±20%, and should also fulfill at least 50% of the calibration standards tested per concentration level	A minimum of six calibration concentration levels exclusive blank and a zero sample	At least 75% of the calibration standards should be ±15% of the nominal value, except for the LLOQ for which it should be within ±20%, and should also fulfill at least 50% of the calibration standards tested per concentration level	FDA (2001)
				For bioequivalence studies, the LLOQ should not be higher than 5% of Cmax	EMA (2011)
QC's	NA	NA	Include QC samples at the following three concentrations (within the calibration range) in duplicate with each analytical batch: Low: near the LLOQ (up to 3× LLOQ) Medium: midrange of calibration curve High: near the high end of range. Each analytical batch should contain six or a minimum of 5% of the total number of unknown samples.	QCs prepared at all concentrations greater than LLOQ <15%	FDA (2001)

			FDA (2001)	FDA (2007)
Accuracy and precision	Analysis of QC samples at four concentration levels: LLOQ, LOW, MEDIUM, and HIGH; six times per analytical batch; and in at least three independent analytical runs. For the within- and between-run accuracy, the mean concentration should be within 15% of the nominal values for the LOW, MEDIUM, and HIGH levels and be within 20% of the nominal value for the LLOQ. The within- and between-run CV value should not exceed 15% for the LOW, MEDIUM, and HIGH levels and not exceed 20% for the LLOQ.	NA	NA	At least 67% (4 of 6) of the QC samples should be within the above limits; 33% of the QC samples (not all replicates at the same concentration) can be outside the limits. If there are more than 2 QC samples at a concentration, then 50% of QC samples at each concentration should pass the above limits of deviation.

(continued overleaf)

Table 11.1 *(continued)*

Validation parameter	Pre-study experiment	Acceptance criteria	In-study experiment	Acceptance criteria	Introduced
Dilution integrity	Dilution integrity should be demonstrated by spiking the matrix with an analyte concentration above the ULOQ and diluting this sample with blank matrix (at least five determinations per dilution factor).	Accuracy and precision should be within the set criteria, that is, within ±15%.	Dilution integrity should cover the dilution applied to the study samples.	Accuracy and precision should be within the set criteria, that is, within ±15%.	EMA (2011)
Selectivity	Selectivity should be proved using at least six individual sources of the appropriate blank matrix, which are individually analyzed and evaluated for interference.	Absence of interfering components is accepted where the response is less than 20% of the lower limit of quantification for the analyte and 5% for the internal standard.	Inject zero and blank as part of the analytical run.	An absence of interfering components is accepted where the response is less than 20% of the lower limit of quantification for the analyte and 5% for the internal standard.	FDA (2001)

			FDA (2007)
Matrix factor	The quantitative measure of matrix effect can be termed as matrix factor (MF) and defined as a ratio of the analyte peak response in the presence of matrix ions to the analyte peak response in the absence of matrix ions. The internal standard (IS)-normalized MF is the MF of analyte divided by the MF for IS. In addition to the normal matrix, it is recommended to investigate matrix effects on other samples, for example hemolyzed and hyperlipidemic plasma samples. If samples from special populations (e.g., renal- or hepatic-impaired populations) are to be analyzed, it is also recommended to study matrix effects using matrix from such populations.	The variability in matrix factors, as measured by the coefficient of variation, should be less than 15%.	NA
	NA	—	EMA (2011)
		—	
Repro-ducibility	Reinjection reproducibility should be evaluated to determine if an analytical run could be reanalyzed in the case of instrument failure.	See calibration line, accuracy, and precision.	FDA (2001)
	—	—	

(continued overleaf)

Table 11.1 *(continued)*

Validation parameter	Pre-study experiment	Acceptance criteria	In-study experiment	Acceptance criteria	Introduced
Incurred sample reproducibility (ISR)	NA	NA	Stability and reproducibility of incurred samples should be addressed for all analytical methods employed. It is left to the investigators to use their scientific judgment to address the issue.		FDA (2007) At least 67% of the repeats should be within 20% of their mean.
			Ten percent of the samples should be reanalyzed in case the number of samples is less than 1000 and 5% of the number of samples exceeds 1000 samples.		EMA (2011)
			The total number of ISR samples should be 7% of the study sample size.		FDA draft (2013)
Carry-over	Inject blank samples after an upper limit of quantification.	The concentration in the blank should not be greater than 20% of the LLOQ.	Inject blank samples after an upper limit of quantification.	The concentration in the blank should not be greater than 20% of the LLOQ.	EMA (2011)

| Long-term frozen | Long-term stability should be determined by storing at least three aliquots of each of the low and high concentrations under the same conditions as the study samples. The concentrations of all the stability samples should be compared to the mean of back-calculated values for the standards at the appropriate concentrations from the first day of long-term stability testing. | NA | NA | FDA (2001) |

(continued overleaf)

Table 11.1 *(continued)*

Validation parameter	Pre-study experiment	Acceptance criteria	In-study experiment	Acceptance criteria	Introduced
	Stability evaluations should be performed against freshly prepared standard curves. When evaluating data generated from stability experiments, intended (nominal) concentrations should be used for comparison purposes.				FDA (2007)
	Stability of the analyte in the studied matrix is evaluated using low and high QC samples (blank matrix spiked with analyte at a concentration of a maximum of three times the LLOQ and close to the ULOQ), which are analyzed immediately after preparation and after the applied storage conditions that are to be evaluated.	The mean concentration at each level should be within ±15% of the nominal concentration.			EMA (2011)

Bench-top	Three aliquots of each of the low and high concentrations should be thawed at room temperature and kept at this temperature from 4 to 24 hours	NA	NA	FDA (2001)
	The mean concentration at each level should be within ±15% of the nominal concentration.			EMA (2011)
Freeze-thaw	At least three aliquots at each of the low and high concentrations should be stored at the intended storage temperature for 24 hours and thawed unassisted at room temperature. When completely thawed, the samples should be refrozen for 12–24 hours under the same conditions.	NA	NA	FDA 2001
	The mean concentration at each level should be within ±15% of the nominal concentration.			EMA 2011

(continued overleaf)

Table 11.1   *(continued)*

Validation parameter	Pre-study experiment	Acceptance criteria	In-study experiment	Acceptance criteria	Introduced
Post-preparative	The stability of processed samples, including the resident time in the auto-sampler, should be determined.		NA	NA	FDA (2001)
		The mean concentration at each level should be within ±15% of the nominal concentration.			EMA (2011)
Stock solution stability	The stability of stock solutions of drug and the internal standard should be evaluated at room temperature for at least 6 hours.		—	—	FDA (2001)
		The mean concentration at each level should be within ±15% of the nominal concentration.			EMA (2011)

LLOQ, lower limit of quantification; NA, not available; QC, quality control.

11.5   In-Study Validation

The setup of a bioanalytical run (batch) should mimic the run during validation. This covers a calibration line identical to the one validated, *blank samples* (samples of a biological matrix to which no analyte has been added), *zero samples* (blanks added with internal standard), and *QC samples* bracketed around the study samples with unknown concentrations (often denoted as unknowns). When the *in-study validation* term was introduced, it was based on validation of accuracy and precision within an analytical run by evaluation of back-calculated standards to nominal concentration and quantification of spiked QC samples, with concentrations evenly distributed over the dynamic range of the method. The statistics and acceptance criteria for this reflected those applied in the pre-study validation. Further selectivity in blanks and zero samples should be monitored.

The reproducibility of the method as an in-process requirement has been added as the requirements to bioanalysis have developed, and it is defined as the ISR. This means to demonstrate assay reproducibility by reanalysis of selected study samples to confirm the original result. ISR should be considered as a process control combined with an in-study validation of the method when applied to study samples, and the reanalysis should be conducted under the exact same conditions as the origin analysis and conducted shortly after the first analysis in order to eliminate stability issues in the bias. ISR should be conducted at least in the toxicokinetic studies once per species, in first clinical trial in subjects, in all pivotal bioequivalence trials, in first patient trial, and in first trial in patients with impaired hepatic and/or renal function. If the method is changed significantly and revalidated, then ISR should be repeated in subsequent studies. A number of publications have provided guidance on how, when, and to what extent the ISR should be conducted, but as a general rule 10% of samples from a given study should be repeated if a study contains analysis of a new population, species, or analyte. Two out of three of the reanalyzed samples should have a result less than 20% from the mean result.

Also, *auto-sampler carry-over* is a pre-study validation parameter that should be evaluated during a study, as carry-over is a dynamic element that may change within an analytical batch as a result of, for example, build-up of matrix residues to the system, or threadbare parts in the auto-sampler. The carry-over should be monitored and measured for every analytical run. Auto-sampler carry-over could be monitored by injection of a blank sample after a ULOQ calibration standard. The carry-over should be evaluated as selectivity, and thus the sum of interaction peaks from matrix and carry-over should not exceed 20% of the peak area of the lowest calibration standard.

The *dilution integrity* is a parameter that should be validated during pre-study validation as well as an in-study validation. However, as the dilution integrity usually covers manual pipetting or robotic pipetting by qualified robots, it is a dynamic process like the carry-over assessment, and it makes sense to conduct the test of dilution integrity during the analytical run. An in-study validation of dilution integrity would usually involve dilution of a dedicated QC sample with the same dilution regime as the most diluted sample in a particular run, for example, not only the dilution factor but also the exact combination of volumes and dilutions to obtain the final dilution of the unknown sample.

11.6   Documentation

As stated in the introduction to this chapter, a method validation should be conducted after the final method optimization as a systematic collection of the documentation that the performance characteristics of the method are suitable and reliable for the intended analytical applications and can meet predefined acceptance criteria for each parameter. This collection of documentation should be straightforward and easy to understand for the authorities, and it is highly recommended to conduct BMV studies for the purpose of regulated bioanalysis as a protocol and/or Standard Operation Procedure (SOP)-driven process and to apply good record keeping to all processes during the validation study. Documentation of successful completion of such a study should be provided in a BMV report.

11.7   Regulatory Requirements to Bioanalysis

Bioanalytical data are an essential and integrated part of the safety evaluation of new drugs. Prior to first dose of a drug under development to humans, the relationships between dose, systemic exposure, and adverse effect in the test animals are studied carefully in order to establish a NOAEL. When administrating a new drug to a human for the first time, and during subsequent dose escalation in the SAD (*single ascending dose*) and MAD (*multiple ascending dose*) studies, the healthy subjects are monitored for a large number of safety markers and biomarkers to ensure the safety of the subjects. Bioanalysis is used for determining systemic exposure of both the drug under investigation and relevant metabolites. The exposures in the human subjects are related to the exposure at NOAEL in the toxicity species, and this relation serves as guidance for a safe maximal exposure to humans, and thus guidance for dose escalation, securing that humans are not exposed to higher concentrations than for which safety evaluation data in test animals exist.

From the bioanalytical scientist's perspective, it would be logical if the requirements to the performance and level of validation of the bioanalytical method could increase with growing experience in analysis of the compound as it proceeds along the drug development process; that is, if the requirements to the performance of the bioanalytical method could be increased as experience is gained and the method is improved (Figure 11.1). However, as per guidance, the method used for exposure determination in the first toxicological studies in the very early days of drug development in the preclinical phase should be as reliable as the exposure determinations in the later clinical phases. This is in order to be able to determine the exposure ratio correctly, as this ratio is not only essential to human safety during clinical development but also an important part of the assessment of a new drug's risk–benefit ratio.

A special case where the importance of bioanalytical method reliability is crucial for human efficacy and safety is in the case of bioequivalence studies (BE) because these studies are pivotal in the approval of new formulations and generic versions of a drug. The commercial implications of the outcome of such a study may be significant, and the BE studies are often under scrutiny from the authorities. The BEs are conducted as cross-over studies (where the subjects serve as their own reference) to eliminate the biological variability, and the applied method should be optimized with a narrow dynamic range reflecting the expected concentrations to optimize the accuracy and precision of the measurements.

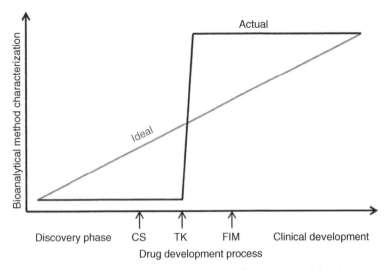

Figure 11.1 *Level of method characterization during the process of development of a new drug. The level of method characterization increases dramatically after the drug candidate has been selected for development and remains hereafter constant from the first regulated toxicological studies in the early preclinical development phase throughout the clinical development phase. CS = candidate selection, TKs = toxicokinetics, and FIM = first in man*

Other studies where bioanalysis is considered pivotal and methods should be fully validated are PK studies where the human pharmacokinetics is investigated under the influence of other drugs, alcohol, or other confounders.

There are, however, also areas where less validation is required. As the biological variability in pharmacokinetics in the populations of test animals or human subjects may be large, the requirements to methods bias and validation requirements may seem unnecessarily strict for studies where, for example, pharmacokinetics and human safety are not the primary outcomes of a study. In the draft FDA BMV guidance from 2013, it is stated, "For pivotal studies that require regulatory action approval or labeling, such as BE or PK studies, the bioanalytical methods should be fully validated. For exploratory methods used for the sponsor's internal decision making, less validation may be sufficient." Also, in 2008, the FDA issued guidance on evaluation of MIST as referred to earlier, and at the same time the FDA invited to apply a tier approach in the characterization of exposure of human metabolites, that is, to apply different levels of validation for the metabolite quantification at each stage of the drug development process. In other words, it is considered sufficient to explore with less characterized methods until the relevant human metabolites are identified, and then eventually provide documentation for the human and animal exposure at relevant doses with guidance-validated methods.

11.8   Quality Systems in Regulated Bioanalysis

The general expectations from the regulators to bioanalytical laboratories are outlined in the FDA BMV guidance from 2001 and in the EMA 2011 guideline. As facilities conducting pharmacology, toxicology, and other preclinical studies for regulatory submissions, bioanalytical laboratories should adhere to GLPs and to sound principles of quality assurance.

GLP is a quality system developed for management control for research laboratories and organizations to ensure consistent and reliable quality of nonclinical safety tests. GLP applies to nonclinical studies conducted for the assessment of the safety or efficacy of chemicals (including pharmaceuticals) to man, animals, and the environment. The original GLP regulatory mandate was promulgated in 1976 by the FDA and a few years later (1981) by the Organization for Economic Cooperation and Development (OECD).

The GLP guideline does not cover bioanalysis in particular. However, when it comes to BMV as an independent bioanalytical study and whether this study should be conducted in the principals of or even according to GLP requirements have been issues of debate for more than 25 years within pharmaceutical companies and the bioanalytical community. Within companies, it has been discussed if additional resources of quality assurance (QA) involvement in the BMV process represented a valuable investment. It was argued that it made little sense that BMV, being an important prerequisite for regulated nonclinical and clinical studies, was not covered by GLP requirements. However, QA involvement in an early phase of drug development method validation could be a rate-limiting step in the overall development process, delaying new drugs becoming available to patients and losing valuable patent exclusivity time for the pharmaceutical companies. Undoubtedly, a wide spectrum of opinions existed on this subject from "better safe than sorry" to "a waste of resources"; and different degrees of QA involvement existed depending on the individual company's experience, with the extremes being no involvement at all and full involvement of QA with approval of validation protocol prior to the start of the experimental phase, audit of the experimental phase and data, and report audit. Although it has been claimed that it was the expectation by authorities that validation of bioanalytical methods (to be used for analyzing samples from GLP regulated studies and regulated clinical studies) also was conducted according to the requirement of GLP, until recently no official requirement existed that BMV should be GLP compliant.

A milestone was the publication of the EMA Guideline on BMV method validation on 21 July 2011 with official legal requirements to the validation of bioanalytical methods: "Normally, the validation of bioanalytical methods used in non-clinical pharmaco-toxicological studies that are carried out in conformity with the provisions related to GLP should be performed following the Principles of Good Laboratory Practice." Further from the EMA Reflection paper on guidance for laboratories that perform the analysis or evaluation of clinical trial samples, published 26 August 2010 (EMA/INS/GCP/532137/2010): "The validation of bioanalytical methods and the analysis of study samples for clinical trials in humans should be performed following the principles of GCP ("Reflection Paper for Laboratories That Perform The Analysis Or Evaluation Of Clinical Trial Samples")." With the clear guidance from EMA since 2011, both pharmaceutical companies and *contract research organizations* (CROs), the latter conducting bioanalysis on behalf of pharmaceutical companies, now conduct BMV according to GLP, and it goes without saying that such an approach requires close coordination and good collaboration between bioanalytical scientists and QA people in order not to become a bottleneck and delay the overall drug development process.

There is a general agreement between bioanalytical and QA people that method validation should be conducted prior to the analysis of actual study samples from regulated nonclinical or clinical studies. It would be unethical not to get an optimal data quality of samples for which subjects have bled and animals have been sacrificed because of a bioanalytical method not performing satisfactorily and not passing predefined objective acceptance criteria. In line with the in-study validation philosophy, the authorities have focused on method performance and the number of failed analytical runs out of the total number of analytical runs in a given study. This may be when the authorities conduct inspections or assess Marketing Authorization Applications in the European Union (MAA) or New Drug Applications in the United States (NDA). A high percentage of failed runs will draw attention to the auditor or assessor, who might question if the method is fit for purpose and ultimately if the data are reliable and suitable for submission. As a company and certainly as being responsible for bioanalysis, that is not a situation you want to be in.

Today, bioanalysis is an extensively regulated area, and the ability to comply with regulatory demands is essential in having new drugs approved. Authorities can for formal reasons reject NDAs in case pivotal nonclinical and/or clinical studies are not conducted according to GLP/GCP. Because unreliable bioanalytical data can affect human safety, violating GLP principles is considered a very serious issue by the authorities. In a recent case from 2013, a bioanalytical scientist employed at an international CRO was sentenced to three months in prison in the United Kingdom for irregularities in bioanalytical data generated to support nonclinical studies as well as clinical trials. The Medicines and Healthcare Products Regulatory Agency (MHRA), which is the UK government agency responsible for ensuring that medicines and medical devices work and are acceptably safe, posted a press release from April 2013 stating, "Man jailed in pre-clinical trial data scam case." In this press release, MHRA writes, "The sentence sends a message that we will not hesitate to persecute those whose actions have the potential to harm public health." Following a full assessment by the MHRA's inspection team and assessors, it was concluded that the data integrity issues did not invalidate the results of the clinical trials that were affected. However, as a result of the actions, the development of a number of new medicines was significant delayed and considerable cost to the study sponsors (pharmaceutical companies) was incurred as a result of the delay.

Many pharmaceutical companies these years choose to outsource bioanalyses from their pivotal studies to CROs, for example for cost or strategic reasons. However, the example given here clearly shows that only the bioanalytical work can be outsourced. The responsibility and the consequences of noncompliance can have serious impact not only for the responsible bioanalytical scientist but also for the pharmaceutical company, which can risk having delays or even rejection of an NDA/MAA due to compliance issues.

Index

Note: page numbers in *italics* indicate figures; those in **bold** tables.

Bioanalysis of Pharmaceuticals: Sample Preparation, Separation Techniques and Mass Spectrometry,
First Edition. Steen Honoré Hansen and Stig Pedersen-Bjergaard.
© 2015 John Wiley & Sons, Ltd. Published 2015 by John Wiley & Sons, Ltd.